THE NEW GROVE®

PIANO

THE NEW GROVE®

DICTIONARY OF MUSICAL INSTRUMENTS

Editor: Stanley Sadie

The Grove Musical Instruments Series

ORGAN

PIANO

in preparation

BRASS

EARLY KEYBOARD INSTRUMENTS

VIOLIN FAMILY

WOODWIND

THE NEW GROVE®

PIANO

Edwin M. Ripin
Philip R. Belt
Maribel Meisel
Derek Adlam
William J. Conner
Rosamond E. M. Harding
Cyril Ehrlich
Hugh Davies
Robert Winter
J. Bradford Robinson
Harold C. Schonberg
Susan Bradshaw

MACMILLAN

Parts of this material first published in
The New Grove® Dictionary of Musical Instruments,
edited by Stanley Sadie, 1984

and

The New Grove Dictionary of Music and Musicians®,
edited by Stanley Sadie, 1980

First published in UK in paperback with additions 1988 by
PAPERMAC
a division of Macmillan Publishers Limited
London and Basingstoke

First published in UK in hardback with additions 1988 by
MACMILLAN LONDON LIMITED
4 Little Essex Street London WC2R 3LF
and Basingstoke

British Library Cataloguing in Publication Data

Piano.—(The new Grove musical instruments series)
1. Piano—History 2. Piano music—History and criticism
I. Belt, Philip R. II. The new Grove dictionary of musical instruments
786.2′1′09 ML650

ISBN 0–333–4447–7 (hardback)
ISBN 0–333–4448–5 (paperback)

First American edition in book form with additions 1988 by
W. W. NORTON & COMPANY
500 Fifth Avenue, New York NY 10110

ISBN 0-393-02553-5 (hardback)
ISBN 0-393-30514-7 (paperback)

Printed in Hong Kong

Contents

List of illustrations

Illustration acknowledgments

We are grateful to the following for permission to reproduce illustrative material: Musikinstrumentenmuseum, Karl-Marx-Universität, Leipzig (fig.2); Bildarchiv Preussischer Kulturbesitz, Berlin (fig.3); Germanisches Nationalmuseum, Nuremberg (figs.6, 7 Rück Collection fig.15 Neupert Collection); Board of Trustees of the Victoria and Albert Museum, London (figs.8, 31); Yale Center for British Art (Paul Mellon Collection), New Haven, Connecticut/ photo Royal Academy of Arts, London (fig.9); Smithsonian Institution, Washington, DC (figs.13, 27, 29*a*); Öffentliche Kunstsammlung, Basle (fig.14); Gesellschaft der Musikfreunde, Vienna/ photo Kunsthistorisches Museum (fig.17); Colt Clavier Collection, Bethersden, Kent (fig.24); Richard Burnett Collection of Historical Keyboard Instruments, Goudhurst (fig.25); Robert Winter (fig.26); Steinway & Sons, New York (figs.28, 29*b*, 38); Ruth Davis, London (fig.30); John Broadwood & Sons Ltd, London (fig.34); L. Bösendorfer Klavierfabrik AG, Vienna/ photo Lucca Chmel (fig.35); Performing Artservices Inc., New York/ photo Remy Charlip (fig.36); Oberfinanzdirektion, Munich/ photo Österreichische Nationalbibliothek (fig.37); Camera Press Ltd, London/ photo Karsh, Ottawa (fig.39); photo Clive Barda, London (figs.40, 41); Board of Trustees of the Victoria and Albert Museum, London (cover).

Music example acknowledgments

We are grateful to the following music publishers for permission to reproduce copyright material: Universal Edition A.G. (Alfred A. Kalmus Ltd) (exx.1, 3); Editions Durand S.A., Paris/United Music Publishers Ltd, London (ex.2).

Abbreviations

AMZ	*Allgemeine musikalische Zeitung*
AMz	*Allgemeine Musik-Zeitung*
BurneyH	C. Burney: *A General History of Music from the Earliest Ages to the Present* (London, 1776–89)
c	circa [about]
CBY	*Current Biography Yearbook*
CMc	*Current Musicology*
D	Deutsch catalogue [Schubert]
DAB	*Dictionary of American Biography*
EMDC	*Encyclopédie de la musique et dictionnaire du Conservatoire*
fl	floruit [he/she flourished]
GSJ	*The Galpin Society Journal*
H	Hoboken catalogue [Haydn]
HMYB	*Hinrichsen's Musical Year Book*
JAMIS	*Journal of the American Musical Instrument Society*
JAMS	*Journal of the American Musicological Society*
K	Köchel catalogue [Mozart]
MJb	*Mozart Jahrbuch des Zentralinstituts für Mozartforschung*
ML	*Music and Letters*
MQ	*The Musical Quarterly*
MR	*The Music Review*
MT	*The Musical Times*
MTNA	*Music Teachers National Association*
PRMA	*Proceedings of the Royal Musical Association*
R	photographic reprint
repr.	reprint
SIMG	*Sammelbände der Internationalen Musik-Gesellschaft*
SMw	*Studien zur Musikwissenschaft*
spr.	spring
suppl.	supplement
WoO	Werke ohne Opuszahl [works without opus number: Beethoven]
WQ	Wotquenne catalogue [C. P. E. Bach]

Preface

This volume is one of a series of short studies derived from *The New Grove Dictionary of Musical Instruments* (London, 1984). Some of the texts were originally written for *The New Grove Dictionary of Music and Musicians* (London, 1980), in the mid-1970s. For the present reprint, all the articles have been reread and modified, mostly by their original authors, and corrections and changes have been made in the light of recent work.

This volume brings together several contributions to the instrument dictionary: the section on repertory, information on individual piano manufacturers (in the Index of Piano Makers), technical information on the components of the piano (in the Glossary of Terms) and an account of the repertory. The section on exponents of the instrument, by Harold C. Schonberg, is entirely new, commissioned specifically for this volume.

I should like particularly to acknowledge the help given by Derek Adlam, who gave generous advice on the Index and the Glossary, and Howard Schott, who made a number of helpful suggestions regarding the script.

S.S

CHAPTER ONE

History of the Piano

1. INTRODUCTION

The piano has occupied a central place in professional and domestic music-making since the third quarter of the 18th century. In addition to the great capacities inherent in the keyboard itself – the ability to sound simultaneously at least as many notes as one has fingers and therefore to be able to produce an approximation of any work in the entire literature of Western music – the piano's capability of playing notes at widely varying degrees of loudness in response to changes in the force with which the keys are struck, permitting crescendos and decrescendos and a natural dynamic shaping of a musical phrase, gave the instrument an enormous advantage over its predecessors, the clavichord and the harpsichord. (Although the clavichord was also capable of dynamic expression in response to changes in touch, its tone was too small to permit it to be used in ensemble music; the harpsichord, on the other hand, had a louder sound but was incapable of producing significant changes in loudness in response to changes in touch.) The capabilities later acquired of sustaining notes at will after the fingers had left the keys (by means of pedals) and of playing far more loudly than was possible on the harpsichord made this advantage even greater.

The instrument's modern name is a shortened form of that given in the first published description of it by Scipione Maffei (1711), where it is called 'gravicembalo col piano e forte' ('harpsichord with soft and loud'). 18th-century English sources used the terms 'pianaforte', 'piano-forte' and 'fortepiano' interchangeably with 'pianoforte'; some scholars reserve 'fortepiano' for the 18th- and early 19th-century instrument, but the cognate is used in Slavonic countries to refer to the modern piano as well. Schubert referred unambiguously to the 'fortepiano' until 1815 (as did Beethoven), but thereafter used the term 'pianoforte'. Responding to the early 19th-century reaction against using foreign words in German, Beethoven and his publisher Steiner preferred the German word 'Hammerclavier' in several of his late sonatas. This term was previously used to distinguish the square piano from the grand piano ('Flügel'); it is now occasionally used for the piano in general (with the spelling 'Hammerklavier') but is usually applied in a historical context.

There is no continuity between the remote 15th-century precursors of the piano (such as the dulce melos, essentially a keyed dulcimer) and the origins of the instrument as discussed below.

The modern piano consists of six major elements: the strings – three for each note down to *B* or *B*♭, then two for each note, except for the extreme bass, with just one; the massive metal frame that supports the enormous tension that the strings impose (approximately 18 tons or 16,400 kg); the soundboard and the bridges which communicate the vibrations of the strings to the soundboard which, in turn, enables these vibrations to be efficiently converted into sound waves, thereby making the sound of the instrument audible; the action, consisting of the keys, the hammers, and the mechanism that impels the hammers towards the strings when the keys are depressed; the wooden case that encloses all of the foregoing; and the pedals, of which the one at the right (the sustaining or damper pedal) acts to undamp all the strings enabling them to vibrate freely regardless of which keys are depressed, while that at the left (the *una corda*) acts to reduce the volume of tone either by moving the hammers sideways so that they strike only two of the three strings provided for each note in the treble and one of the two strings provided for each note in the tenor, or by bringing the hammers closer to the strings, thus shortening their stroke (the middle or *sostenuto* pedal, when present, acts to keep the dampers raised on only those notes being played at the moment the pedal is depressed) or – on some upright pianos – interposes a strip of cloth between the hammers and the strings to produce a muffled tone. (Traditionally, but inaccurately, the right and left pedals have been called 'loud' and 'soft'.)

Logically, the ideal form of the piano is the 'grand', the wing-shape of which is determined by the fact that the strings gradually lengthen from the treble at the right to the bass at the left. Theoretically, the length of the strings might be doubled for each octave of the instrument's range, but this would be impractical for an instrument having a range of over seven octaves, as the modern piano does, and even the earliest pianos with a range of only four octaves employed some shortening of the strings in the extreme bass. The rectangular 'square' piano, which like the grand has its strings in a horizontal plane, and which was popular in the 18th and 19th centuries, has been entirely superseded by various types of 'vertical' or 'upright' piano, which have their strings in a vertical plane; the fact that uprights take up less room outweighs the disadvantage imposed by the more complex action they must use.

2. ORIGINS TO 1750

The musical advantages initially possessed by the piano were not generally recognized at the time of its invention even though the instrument made its first appearance in a highly developed form, the work of a single individual, Bartolomeo Cristofori, keeper of instruments at the Medici court in Florence. Despite warmly argued claims on the part of such other men as Christoph Gottlieb Schröter and Jean Marius, who experimented with actions for the tangent piano (whose strings are struck by freely moving slips of wood resembling harpsichord jacks rather than by hinged or pivoted jacks, as in the piano), there now seems to be no doubt that Cristofori had actually constructed a working piano before any other maker was even experimenting in this field. According to an entry in Francesco Mannucci's diary, for February 1711, Cristofori had begun work on 'the new harpsichord with piano e forte . . . two years before the Centenary', in 1698; and the detailed description of an 'arpicembalo' in an inventory of the Medici instruments for 1700 establishes that he had by then already completed at least one instrument of this kind. Cristofori's accomplishment as seen in the three surviving pianos made by him, all of which date from the 1720s, would be difficult to exaggerate. His grasp of the essential problems involved in creating a keyboard instrument that sounded by means of strings struck by hammers was so complete that his action included features meeting every challenge that would be posed to designers of pianos for well over a century. Unfortunately, the very completeness of his design resulted in a complicated mechanism, which builders were apparently unwilling to duplicate if they could possibly devise anything that would work and at the same time be simpler to make. As a result, much of the history of the 18th-century piano is the history of the gradual reinvention or readoption of things that were an integral part of Cristofori's original conception; and it was only with the introduction in the 19th century of increasingly massive hammers that the principles discovered by Cristofori could no longer provide the basis for a completely satisfactory piano action, requiring the still more complicated mechanism known today.

The essential difficulty in creating a workable instrument in which

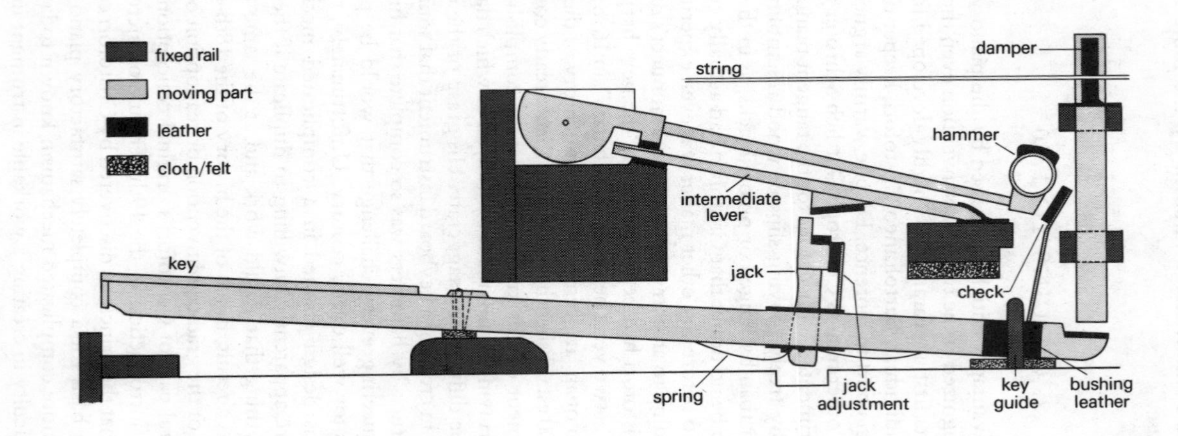

1. Action of a piano by Bartolomeo Cristofori, 1726 (see also fig. 2)

the strings are to be struck by hammers is to provide a means whereby the hammers will strike the strings at high speed and immediately rebound, so that the hammers will not damp out the vibrations they initiate. In order to provide for immediate rebounding, the strings must not be lifted by the impact of the hammer and must therefore be thicker and at higher tension than those of a harpsichord, and the hammers must be tossed towards the strings and be allowed to fly freely for at least some small part of their travel. The smaller this distance of free flight is, the more control the pianist has over the speed with which the hammer will strike the string and accordingly over the loudness of the sound that it will produce. Unfortunately, the smaller this distance is, the more likely the hammer is to jam against the strings or bounce back and forth between the strings and whatever device impelled it upwards when the key was struck; hence when the distance of free flight is made small to permit control of loudness, the hammer is likely to jam or bounce and damp out the tone. Cristofori solved this problem with a mechanism that enabled the hammer to be brought quite close to the string but caused it to fall quite far away from it even if the key was still held down. Devices of this kind are called escapements and they lie at the heart of all advanced piano actions. In addition, Cristofori provided a lever system that caused the hammers to move at a high speed, and a 'check' (or 'back check') which would catch the hammer after it fell so as to eliminate all chance of its bouncing back up to restrike the strings. Finally, his action provided for silencing the strings when the keys were not held down, using slips of wood resembling harpsichord jacks which carried dampers and which rested on the ends of the keys.

These features are all visible in fig.1, which shows the action of the piano of 1726 (see fig.2, p.6) in the Musikinstrumenten-Museum of the Karl-Marx-Universität in Leipzig. When the key is depressed, the pivoted jack mortised through it pushes upwards onto a triangular block attached to the underside of the intermediate lever, which in turn bears on the hammer shank near its point of attachment, providing for a great velocity advantage. (Although the jack rises only about as fast as the front of the key is depressed, the free end of the intermediate lever rises approximately twice as fast and the hammer rises four times more rapidly still.) The escapement is provided by the pivoted jack, which tilts forward just before the hammer reaches the string so that, when the hammer rebounds, the block on the underside of the intermediate lever contacts the padded step at the back of the jack rather than the tip of the jack. As a result, even if the key is held down, the hammer falls to a point at least 1 cm below the strings.

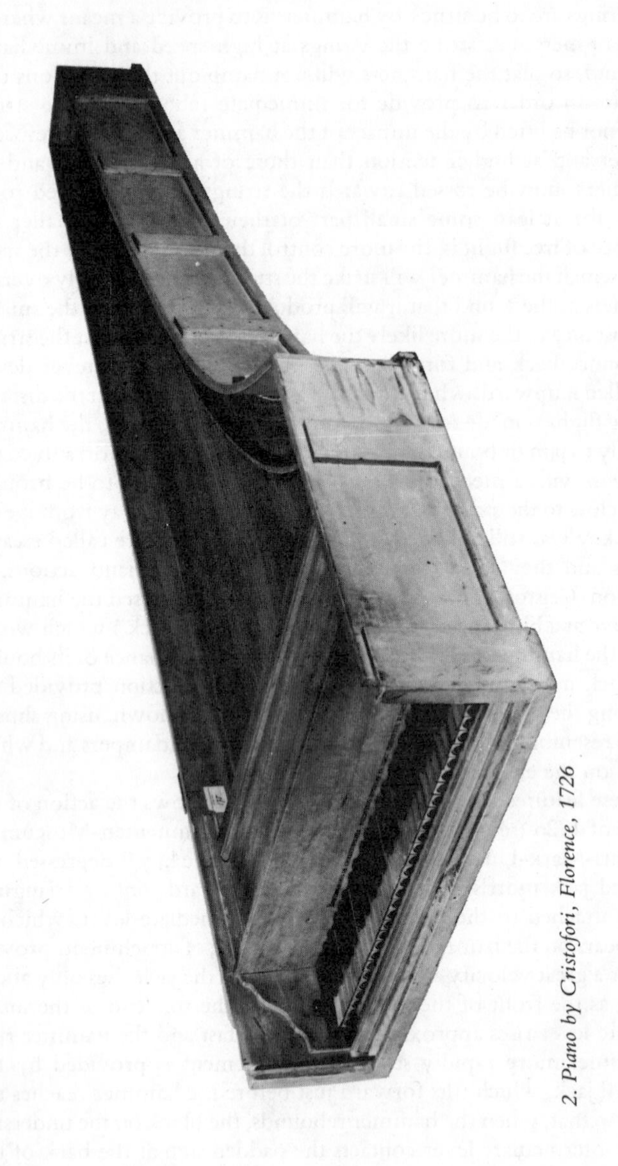

2. Piano by Cristofori, Florence, 1726

Adjustment of the point at which the jack tilts forward is achieved by bending the wire that supports the pad against which the jack is held by the spring. The further forward this pad is, the earlier the jack slips away from the block and the sooner escapement takes place.

The construction of Cristofori's pianos is similar to that of an 18th-century Italian harpsichord of the thick-cased type, except that it is stronger and heavier – for Cristofori obviously recognized the necessity of using thicker strings at higher tension. Thus, the gap between the pinblock or wrest plank and the belly rail (the stout transverse brace that supports the front edge of the soundboard) through which the hammers rise to strike the strings is bridged by a series of wooden braces ('gap spacers') not found in Italian harpsichords. These braces contribute to preventing the wrest plank from twisting or bending into the gap at its centre and are therefore of vital importance in keeping the entire structure from twisting out of shape or collapsing. (The means of ensuring the straightness and integrity of the wrest plank and case structure continued to be one of the principal concerns of piano makers throughout the 18th century.)

Two of Cristofori's three surviving pianos have an inverted wrest plank in which the tuning-pins are driven completely through, with the strings attached to their lower ends after passing across a nut attached to the underside of the wrest plank. According to Maffei, this plan was adopted to provide more space for the action, but it provides at least two other advantages: since the strings bear upwards against the nut, the blow of the hammer, instead of tending to dislodge them, upsetting the tuning and adversely affecting the tone, seats them even more firmly; second, the inverted wrest plank permits placement of the strings close to the top of the action, so that the hammers need not be tall to reach the strings. They can therefore be quite light, an important factor, since Cristofori's lever system, providing for an acceleration of the hammer to eight times the velocity with which the key is depressed, automatically causes the player to feel (at the key) the weight of the hammer multiplied eightfold.

The sound of the surviving Cristofori pianos is very reminiscent of that of the harpsichord owing to the thinness of the strings compared with later instruments and the hardness of the hammers; but it is less brilliant and rather less loud than that of a firmly quilled Italian harpsichord of the time. These points are mentioned in Maffei's account as reasons for the lack of universal praise for the instrument, as is the fact that contemporary keyboard players found the touch difficult to master (in Germany, where the clavichord was used as both a teaching and a practice instrument, no such objection seems to have been raised when the piano became known). On two of the surviving Cristofori pianos it is possible to slide the keyboard sideways so that the hammers strike only one of the two strings provided for each note (the strings are evenly spaced). Apart from this *una corda* capability, Cristofori's pianos make no provision for alteration of the tone by stops or other such devices; however, one would not expect to find such a provision in view of the lack of any multiplicity of stops in Italian harpsichords.

There seems to have been little direct result in Italy of Cristofori's monumental achievement. Maffei, in his account, clearly recognized the important differences between Cristofori's pianos and the harpsichord (even if he had no better name for the new instrument than 'harpsichord with soft and loud'), and an interesting collection of 12 sonatas for the instrument that includes dynamic markings implying crescendos and decrescendos was published in 1732 (Lodovico Giustini's *Sonate da cimbalo di piano e forte*). But only a handful of other Italian instrument makers seem to have followed in Cristofori's footsteps. It was left primarily to German builders and musicians to exploit his work in the years after his death in 1731.

A German translation of Maffei's account was published in Johann Mattheson's *Critica musica* (1722–5) where it was presumably seen by Gottfried Silbermann, who is reported to have begun experimenting on pianos of his own in the 1730s. He is said to have offered one for Bach's inspection, and at the composer's adverse reaction to its heavy touch and weak treble to have gone on to further experiments resulting in improved instruments, a number of which were bought by Frederick the Great. These are reported to have met with Bach's complete approval when he visited Potsdam in 1747, and the composer served as a sales agent for Silbermann in 1749 (see C. Wolff: 'New Research on Bach's *Musical Offering*', *MQ*, lvii (1971), 403). The two Silbermann pianos owned by Frederick that have survived have actions identical with those in the surviving Cristofori instruments; it seems more than likely that by the time Silbermann made them he had seen an example, whereas his earlier attempts had failed as a result of having been based on the diagram accompanying Maffei's description – which Maffei admitted had been drawn from memory without the instrument before him. Silbermann retained Cristofori's inverted wrest plank and the equidistant spacing of the strings (see fig.3, pp.10–11) and he used the hollow hammers made of rolled parchment found in the 1726 instrument which, together with the check replacing silk strands, evidently replaced the small blocks shown in Maffei's diagram. As might be expected from a representative of the north European keyboard instrument building tradition, Silbermann included hand stops for raising the treble and bass dampers in addition to devices for sliding the keyboard sideways so that the hammers would strike only one of the two strings provided for each note. Thus, these two most characteristic means of modifying the piano's tone, integral to all modern pianos, were found together as early as the 1740s.

Although Gottfried Silbermann and his nephew Johann Heinrich Silbermann seem to have made direct copies of Cristofori's instruments virtually unchanged except for the addition of damper-lifting mechanisms, other German makers, some of whom may perhaps not even have been explicitly informed of Cristofori's work to the extent of knowing of the existence of 'hammer harpsichords', devised a host of less complicated actions, many adapted to the rectangular clavichord-shaped square pianos, the earliest surviving example of which was made by Johann Socher in 1742. Socher's piano contains an action of extraordinary simplicity in which a hammer hinged to the back of the case is thrust upwards by a block at the end of the key, reducing Cristofori's mechanism to an absolute minimum. This type of action became known as the *Stossmechanik* and is the principle upon which the later English builders and their followers built their pianos (see §4 below). Socher's instrument included a device for lifting all the dampers, which permitted the strings to vibrate freely like those of a dulcimer: it seems possible that this is the only tangible influence on the early piano of Pantaleon Hebenstreit, the touring dulcimer virtuoso whose name has become inextricably entwined with the early history of the piano owing to the supposed influence of his instrument on Marius in Paris and the early German makers; this is evidenced by the fact that one of the common German names for the early square piano was 'pantaleon' or 'pantalon' (the name that Hebenstreit had given to his elaborate dulcimer, allegedly with the approval of Louis XIV, in 1705).

The great period of piano building in the German-speaking world is not, however, represented by these developments or even by Silbermann's work, which with the death of his son seems to have led to no direct line of Cristofori-inspired instrument building. Rather, a different approach evolved – using a type of action known as the *Prellmechanik* – which dominated German piano building for the next 75 years.

3. Piano by Johann Heinrich Silbermann, Strasbourg, 1766: (a) general view; (b) plan view

(a)

(b)

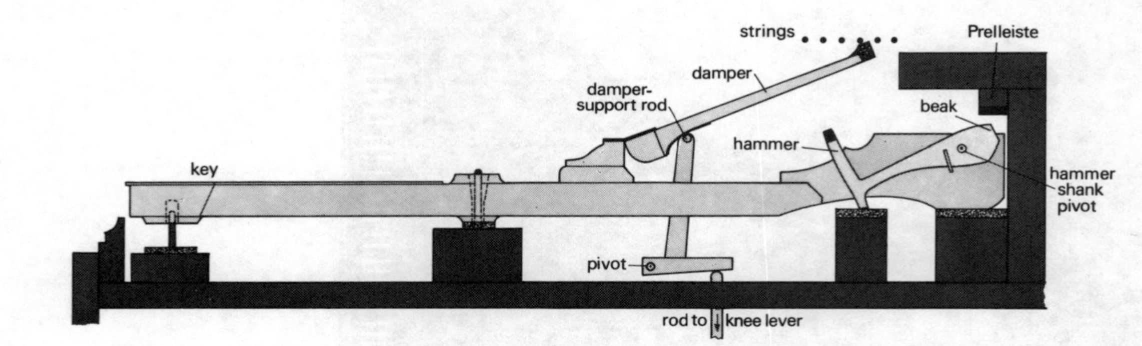

4. Prellmechanik action without escapement from an anonymous south German square piano, c1770; to allow space for the hammer, the rear of the key is narrower than the front; the dampers are beneath the strings and disengage when the hinged end rises with the key; the knee lever lowers the damper-support rod to disengage all the dampers simultaneously

3. GERMANY AND AUSTRIA, 1750–1800

Whereas Cristofori, the Silbermanns and the later piano makers of other schools sought to create a harpsichord capable of dynamic expression, the main thrust of German and Austrian piano building in the later part of the century seems to have been towards creating an instrument that would be like a louder clavichord (Germany and Scandinavia being virtually the only countries in which the clavichord was still esteemed at this period). These German and Austrian pianos have a relatively clear singing tone and an extremely light touch. The simplest of the square models with the *Prellmechanik* show clearly the inspiration of their origin: all that separates them from the clavichord is the addition of a nut at the rear to determine the speaking length of the strings, and the replacement of the tangent with a hammer hinged to the back of the key. In the simple *Prellmechanik* most commonly (and apparently exclusively) used in square models, each of the hammer shanks is attached to its own key – either directly to the top or side (see fig.4), or by a wooden or metal fork or block (the *Kapsel*) – with the hammer head towards the player. A point (the 'beak') on the opposite end of the hammer shank extends beyond the end of the key. This beak is stopped vertically either by the underside of the hitch-pin apron or by a fixed rail called the *Prelleiste*: as the back of the key rises, the hammer is thereby flipped upwards towards the string. As the distance from the tip of the beak to the hammer shank pivot is far shorter than the distance from the pivot to the hammer, the hammer ascends much more rapidly than does the back of the key. An adequate free-flight distance had to be left as there was no escapement to prevent the hammers from restriking the string or blocking and interrupting the tone.

The development of an individual escapement for the *Prellmechanik* has often been credited to Johann Andreas Stein (1728–92), a keyboard instrument maker in Augsburg; but the history of the earliest pianos having this improvement is obscure because of a general lack of documentation and because labels may be missing, inaccurate or sometimes falsified, often failing to agree with dated signatures inside the instruments themselves. An instrument by Stein, labelled 1773 (though the date may not be correct) and now in the Karl-Marx-Universität, Leipzig, may well be the oldest extant piano with the *Prellmechanik* escapement. By 1777 that type of action must have evolved sufficiently to satisfy Mozart when he visited Stein in Augsburg (Mozart complained of hammers blocking and stuttering on other instruments), although the Verona harpischord-piano of the same year has a stationary

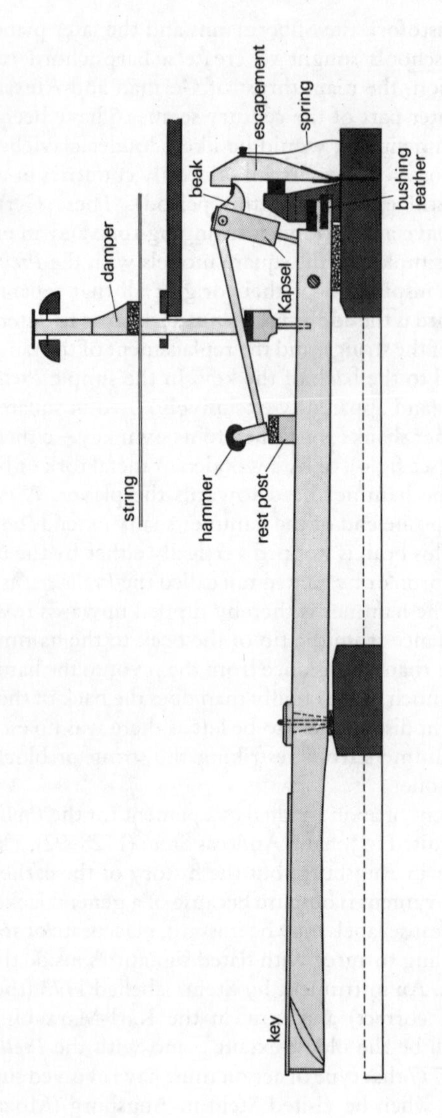

5. Prellmechanik with escapement, believed to have been first used by Johann Andreas Stein, from a Heilman piano, c1785

mechanism while the individual escapement levers are activated by the keys (*Zuggetriebe*; see Pfeiffer, 1948).

In the developed *Prellmechanik* there are escapement levers for each key instead of a stationary rail. Each lever has a notch into which the beak of the hammer shank fits, and each lever has its own return spring (see fig.5). As the key is depressed, the beak is caught by the top of the notch in the escapement lever, lifting the hammer. The combined arcs traversed by the key and the hammer shank cause the beak to withdraw from the escapement and slip free just before the hammer meets the string, after which it is free to fall back to its rest position. When the key is released, the beak slides down the face of the escapement lever back into the notch.

An important feature in such pianos is the extremely small and light hammers (see fig.19, p.37); their thin leather covering (instead of felt) is vital to these instruments' clavichord-like delicacy of articulation and nuance. Typically, the Stein action has either round hollow hammers similar to those of the Silbermanns but made of hazelwood, or short solid hammers usually made of pearwood (the *Kapseln* are also of pearwood). Surviving Stein instruments from 1781 to 1783 all have the round hollow hammers, as do the instruments of J. D. Schiedmayer (1753–1805), who worked for Stein from 1778 to 1781. In Stein's instruments each key has a post supporting the hammer in a rest position above the level of the keys; this rest post is covered with a soft thick cloth which helps absorb the shock of the returning hammers as there are no back checks. To place the action in its proper position (behind the wrest plank in a grand) a 'sled' or drawer about 5 cm deep is slipped under the keyboard to bring the hammer mechanism close to the strings. The keyboard itself is generally of spruce with ebony key facings for the naturals and with sharps of dyed pearwood topped with bone or ivory.

The individual dampers are fitted into a rack above the strings, which the player can raise by means of knee levers under the keyboard. Some of Stein's combination instruments (piano-organ etc) have hand stops and register changes, but these are apparently not original on the fortepiano. The Stein case construction (see figs.6 and 7, pp.16–17) has a double curved bentside with the liner (the rim of the inner frame which supports the soundboard) made of solid blocks of wood sawn to shape. The frame is braced by two or three members perpendicular to the spine (the straight side of the instrument) and several diagonal supports. The case is closed at the bottom by a thick board with the grain running parallel to the straight part of the bentside, and is usually veneered in plain walnut or cherry with a band of moulding around

6. Piano by Stein, Augsburg, 1788 (see also fig.7)

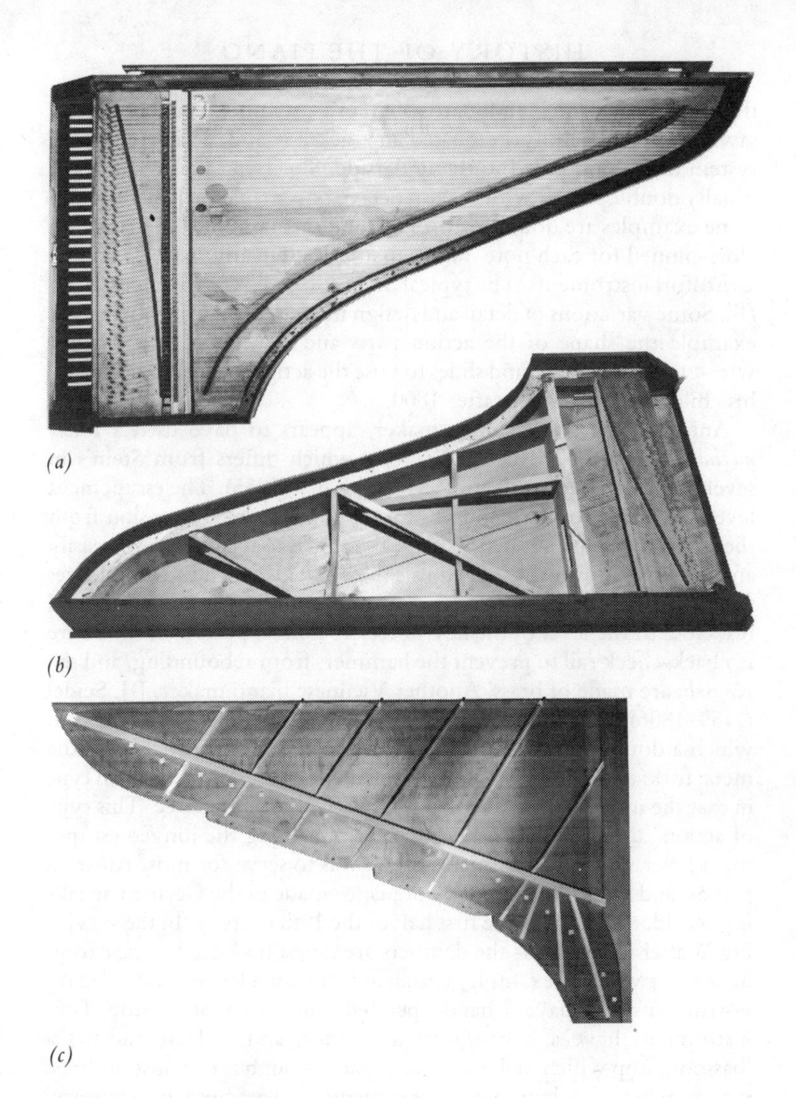

7. Piano by Stein, Augsburg, 1788 (fig.6): (a) plan view; (b) internal structure from above, with soundboard removed; (c) underside of soundboard

the lower edge. The soundboards of Stein's instruments are of quarter-sawn spruce, carefully graduated in thickness and with a complex system of ribbing glued to the underside. Surviving Stein pianos are usually double strung with the top octave or so triple strung, although some examples are double or triple strung throughout; the strings are close-pinned for each note, rather than being equally spaced as in the Cristofori instruments. The typical Stein compass is five octaves, F' to f'''. Some variations of detail and design in Stein's late instruments, for example the shape of the action parts and the use of gap spacers, wire-guided dampers, and slides to raise the action, were continued by his children until shortly after 1800.

Anton Walter, a Viennese maker, appears to have used a *Prellmechanik*, possibly in the early 1780s, which differs from Stein's in several respects (see Luithlen, 1954, and Rück, 1955). The escapement levers are tilted forward and so bear a lighter degree of tension from the escapement spring; hence the beak returns to its notch more easily and quickly. A movable rail adjusts the point at which the beak leaves the escapement lever notch. The hammers are longer and larger and rest close to the level of the key; there are no rest posts as such. There is a back-check rail to prevent the hammers from rebounding, and the *Kapseln* are made of brass. Another Viennese piano maker, J. J. Seidel (1759–1806), is credited with the invention of the brass *Kapsel*, in which a double-pointed axle fits into two tiny dimple sockets in the metal fork: these *Kapseln* have the advantage over Stein's wooden type in that the movement of the shank is relatively friction-free. This type of action, altered only to the extent of providing the hinged escapement lever with an adjustment screw, was to serve for most Austrian pianos, and indeed the majority of pianos made in the German-speaking world, throughout the first half of the 19th century. In the surviving Walter instruments the dampers are raised by knee levers, except in one or two early examples with hand-operated levers. Several early instruments also have a hand-operated *sourdine* or mute stop; later instruments have a knee-operated *sourdine*, and at least one has a 'bassoon' stop which makes a sheet of parchment buzz against the brass strings when the hammers strike them: a characteristic 'janissary' device of the time. Structurally, Walter's instruments appear to have been built up on the bottom (like the Italian harpsichord) and were of two types: the single and the one and a half curved bentside; the compass of his early instruments was either five octaves, F' to f''', or a slightly extended range, F' to g'''. The instruments are usually double strung except for the top octave or so which is triple strung. As all of Walter's instruments are undated and perhaps many unsigned, it is difficult to know without further research exactly how many he made and when the important innovations attributed to him took place.

A number of other makers of pianos with the *Prellmechanik* worked in Germany and Austria before 1800. Most of the surviving instruments demonstrate little change from the basic models of Stein and Walter, but there was some diversity of case construction and action design. Often there are two knee levers which give the player the choice of sustaining the bass or treble separately; or in some pianos the option is to sustain either the treble or the whole. This is especially useful in music with a singing melody in one hand and an accompaniment figure in the other. The followers of Stein, Walter and other makers of the period included Stein's children, Nannette Stein (later Streicher) and Matthäus Andreas (André) Stein (in Vienna from 1794), Johann Schmidt of Salzburg, the younger J. L. Dulcken of Munich, Karl F. W. Lemme of Brunswick, J. J. Könnicke of Brunswick (and, from 1790, Vienna), Johann Schantz and Ferdinand Hofmann of Vienna, and the Schiedmayers of Erlangen and Stuttgart. Vienna eventually became the centre for piano makers using the developed *Prellmechanik*, which accordingly became known, during the 19th century, as the 'Viennese action'.

4. ENGLAND AND FRANCE TO 1800

The piano was known in London by the mid-1750s. The first to reach England was said by Burney (in 'Harpsichord', *Rees's Cyclopaedia*) to have been built by Father Wood, an English monk in Rome, for Samuel Crisp, who purchased it about 1752 during his travels in Italy. Burney's comments show that England was musically ripe for the new instrument: he found its tone 'superior to that produced by quills, with the additional power of producing all the shades of piano & forte by the finger'; even though the touch and mechanism of Wood's instrument were rather crude (nothing quick could be played on it), the 'Dead March' from *Saul* and 'other solemn and pathetic strains, when executed with taste and feeling . . . excited equal wonder and delight'. In a letter of 27 June 1755 the Rev. William Mason wrote to Thomas Gray: 'I bought at Hamburg such a piano forte, and so cheap, it is a harpsichord too of 2 Unisons & the Jacks serve as mutes (when the Piano Forte is play'd) by the cleverest mechanism imaginable. Won't you buy my Kirkman?' Evidence of a

8. Square piano by Johannes Zumpe, London, 1767

piano commercially available in England can be found in Thomas Mortimer's *The Universal Director* (London, 1763) where Frederic Neubauer advertised 'pianofortes, lyrichords, and claffichords' in addition to the familiar harpsichords.

The first important piano maker in England was Johannes Zumpe, who emigrated from Saxony about 1760. He is said to have been a former follower of Gottfried Silbermann, and one of that group of refugee keyboard instrument makers from Germany known as the '12 apostles'. Zumpe's pianos (see fig.8), the earliest surviving example of which (formerly in the Broadwood collection; now in the Württembergisches Landesmuseum, Stuttgart) dates from 1766, filled important social and commercial as well as musical needs in late 18th-century England. J. C. Bach, who had gone to London in 1762, did much to promote the new instrument, performing on 2 June 1768 the first solo piano pieces to be heard at a concert in England. The piano had first been used the previous year to accompany a singer at a public concert and on 19 May 1768 Henry Walsh played a pianoforte solo in Dublin (see Pleasants, 1985). (The earliest known use of a piano is reported from Vienna in 1763, in a 'concert' by Johann Baptist Schmid; see Badura-Skoda, 1980.) Bach was appointed the queen's official private teacher and directed London's most expensive non-theatrical musical events. That was enough to make the little Zumpe pianos fashionable; the socially ambitious could be sure that they were correct. They were relatively simple to make but louder and brighter than a clavichord and more capable of dynamic nuance and gallantry than a spinet. By 1767 Johannes Pohlmann had begun to fulfil orders that Zumpe could not meet, and by the late 1770s hundreds of square pianos based on Zumpe's pattern were being made by native as well as immigrant builders. Zumpe's first piano action, known as the English single action (see fig.10, p.23), reduced that used by Cristofori and Silbermann to the barest essentials in the interests of simplicity and economy. Zumpe substituted for the jack and intermediate lever in Cristofori's action a wire (the pilot) mounted on the key with a leather-covered button at its upper extremity which acted directly upon the hammer. The absence of both an escapement and a back check leaves nothing to prevent the hammer from rebounding to restrike the strings while the key is depressed. Zumpe's and Pohlmann's pianos, as well as the earliest Broadwood squares, use a sprung damper-lever hinged to the back of the piano case above the strings. The damper is raised by a thin wooden rod (the sticker) that passes between the strings and rises when the key is depressed; the damper spring (often made of whalebone or a similar material) expedites its return once the

9. Zumpe-style square piano and a cello: painting, 'The Cowper and Gore Families' (1775), by Johann Zoffany

10. Zumpe-style single action from a square piano by Jacob and Abraham Kirckman, 1775

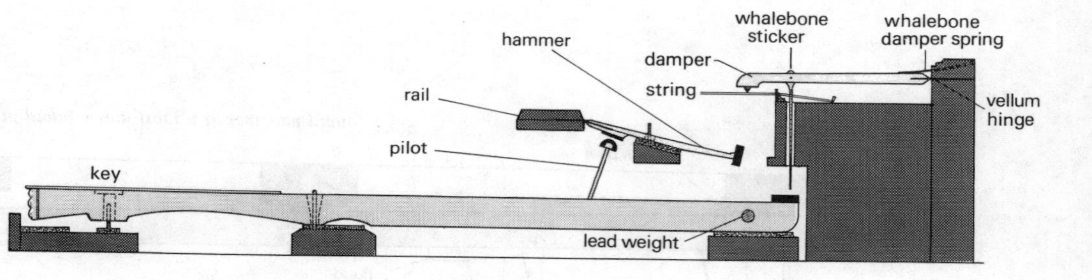

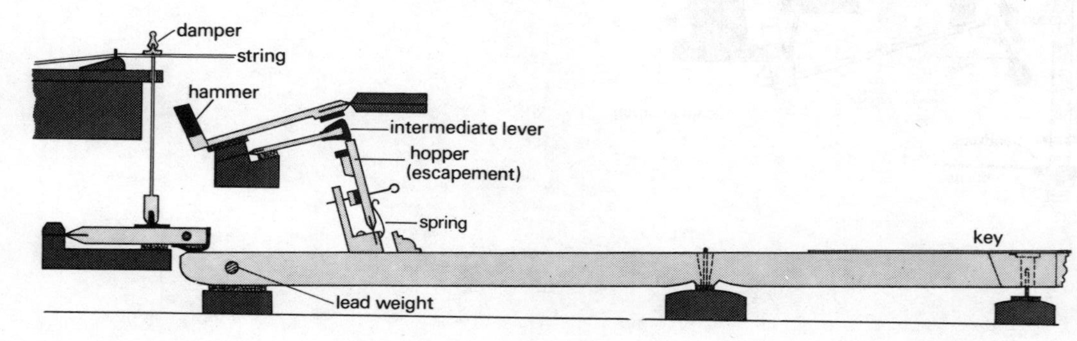

11. Geib-type double action from a piano by Broderip & Wilkinson, c1800

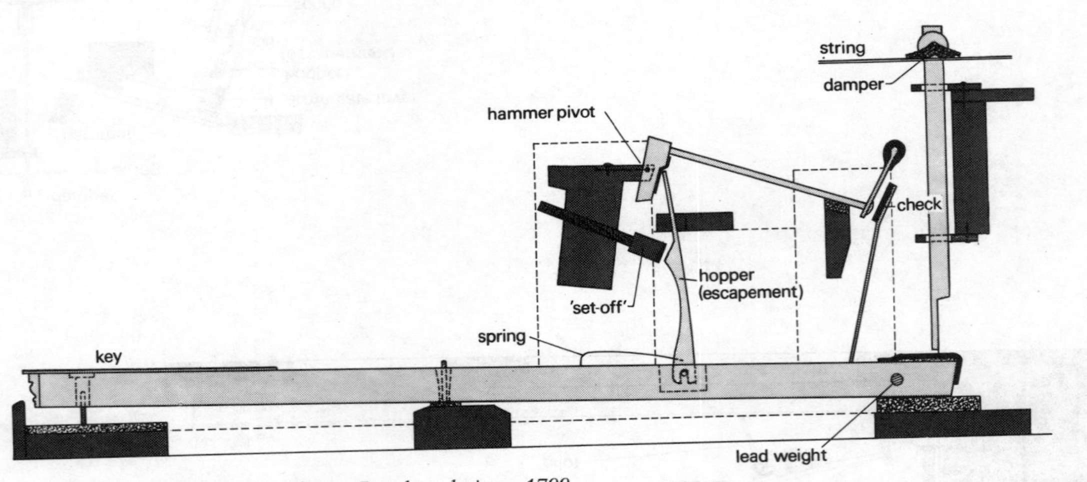

12. English grand piano action from a Broadwood piano, 1799

key is released. During the 1780s some makers began to use a different type, the crank damper, which returns to the string by gravity; instead of the sticker a stiff wire is bent at the top to carry a little wooden head fitted with the damping cloth. This mechanism, more efficient than the lever damper and conducive to a smoother touch, is similar in principle to modern damper mechanisms.

In the 1780s Zumpe developed a double action with an intermediate lever and two pilots. This action was also used by Schoene & Co., successors to Zumpe, as well as Freudenthaler of Paris, Ermel of Brussels, and Erard. A patent for a better double action (see fig.11, p.23) was obtained in 1786 (no.1571) by John Geib, an employee of Longman & Broderip. This action, which includes an escapement in the form of a sprung hopper, is of some sophistication and proficiency – not by any means primitive; eventually it entirely superseded the single action in English square pianos and was also adapted for the vertical 'cabinet' pianos ubiquitous in England from the 1820s to the 1860s.

The English grand piano action was first developed by Americus Backers – with help from John Broadwood and Broadwood's apprentice Robert Stodart – between 1772 and 1776 (later, according to accounts by Broadwood's son James Shudi Broadwood). The earliest surviving piano with the Backers action is an instrument of 1772 now in the Russell Collection, Edinburgh. In the English grand action the intermediate lever is omitted (as in the first Zumpe action) and an escapement or hopper acts directly against the hammer butt; normally the dampers were glued to a slip of wood. Robert Stodart first patented the action in 1777 (no.1172) in his combined piano and harpsichord; the Broadwood firm subsequently adopted and improved it (see fig.12).

The Stodart and Broadwood grands were the first musically important pianos to be built in England. John Broadwood had begun building square pianos in the early 1770s after serving as an apprentice to the harpsichord maker Burkat Shudi; his first grand was built probably in 1781 and the earliest surviving example is from 1786. Among Broadwood's more important improvements in design were his attempts to make string tension throughout the instrument more equal and to have all the hammers striking at a carefully determined point along the string's length. Having called upon the scientific aid of Tiberius Cavallo and of Dr Edward Whitaker Gray of the British Museum, Broadwood divided the bridge on his grand pianos so that the highest brass strings in the bass – thus rendered shorter than the longest steel strings on the treble portion of the bridge – were held at the same tension as the steel; and since a compromise in scaling (to prevent the highest of

(a)

(b)

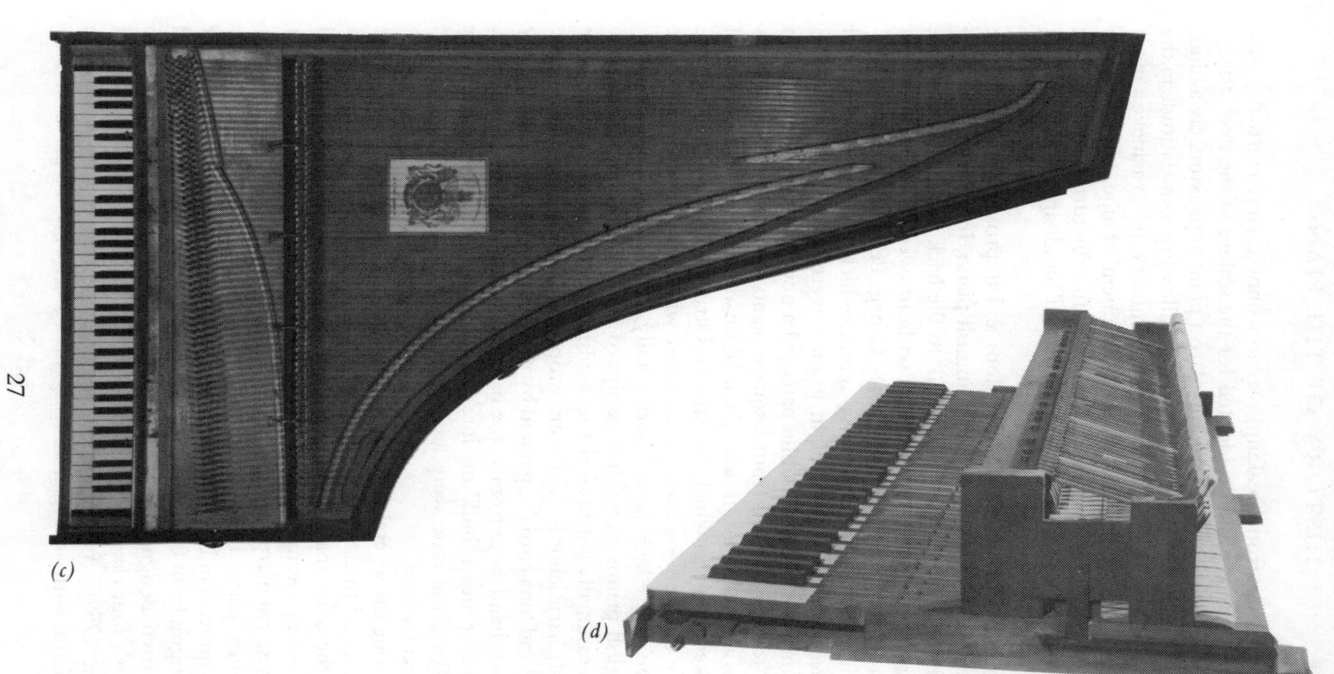

13. Grand piano by John Broadwood & Son, London, 1794: (a) general view; (b) detail of internal bracing showing a rather heavier structure than that in fig.7; (c) plan view; (d) action

the brass strings from breaking) was now unnecessary, a virtually pure harmonic curve could be adopted for the scaling of the steel strings. Broadwood chose as the striking-place a point approximately a ninth along the vibrating length of the string (allowing some latitude in the treble) to suppress certain unwanted harmonics and to achieve a more even tone quality throughout the instrument's range. In appearance the Broadwood grand piano cases until after the turn of the century were like the English harpsichord cases of Shudi and Kirckman (see fig.13, pp.26–7).

A number of additional inventions and improvements was made before 1800 on various kinds of English piano. In his patent of 1783 Broadwood moved the wrest plank to the back of the square piano, giving room for fixing the strings without overcrowding, and for a more regular pressure on the bridges. George Pether of London invented an ingenious down-striking action, which is in fact an inverted *Prellmechanik*. Various 'upright grands' were patented, and in 1798 William Southwell of Dublin patented a new type of upright which seems to have derived from a square piano placed lengthways on its side (with the hitch-pin side upwards) on a stand; for this he invented a 'sticker action' in which the hammers struck the strings (through the agency of extension rods) at the point where they would have done had the square been placed horizontally.

English pianos were often equipped with stops to vary their tone and occasionally with a swell mechanism similar to those found on late English harpsichords. The usual stops on the grand piano were the 'forte' and 'una corda', operated by pedals. Early square pianos usually had two hand stops to raise the bass and treble dampers respectively, and sometimes a 'buff' or 'lute' stop. Apart from these, devices for modifying the tone were less common on English pianos than on continental ones.

Among the several dozen piano makers in London around 1800 the Broadwood firm was pre-eminent, making about 400 instruments a year; the other makers (including the successors to Zumpe, who had returned to Germany in 1783 or 1784) were producing, like their Viennese counterparts, fewer than 40 each.

French piano building at this time was strongly influenced by developments in England, though carried out to a considerable extent by émigrés from Germany. Some French makers had earlier presented their own designs for pianos to the Académie des Sciences (Jean Marius's four *clavecins à maillet*, 1716; Weltmann's modified tangent action, 1759). Adrien l'Epine in 1772 submitted a piano to the Académie with an organ attachment. Pascal Taskin's piano of 1787 (now in the instrument collection of the Stiftung Preussischer Kulturbesitz, Berlin) has a curious action in which the jack for propelling the hammer to the string is fixed to the hammer and not to the key, and the wrest pins are replaced by threaded hooks mounted horizontally in the wrest plank and actuated by a nut. The two halves of a single string of double the required total length are tuned by turning the nut, an experimental system later revived by other French piano makers. Of lasting importance were the activities of Sébastien Erard, who is said to have made his first piano in 1777. He began by copying Zumpe's single action for his square pianos, but by about 1790 he developed his own action *à double pilote*, a modification of Zumpe's second action; and his first grand piano (1796) used a modified English grand action. He also experimented with the *Prellmechanik*, but the main contribution of Erard and his successors was in subsequent technical contributions to the development of the English action.

5. THE VIENNESE PIANO FROM 1800

Of the 200 or so Viennese instrument makers listed in Haupt's study (1960) for the period 1791–1815, at least 135 were keyboard instrument builders. Most prominent were: Anton Walter (1752–1826), who from about 1817 to 1824 was in partnership with his stepson Joseph Schöffstoss (1767–1824); Johann Schantz (*c*1762–1828), who had taken over the workshop of his deceased brother Wenzel in 1791, and whose business was continued from 1831 by Joseph Angst (*c*1786–1842); and Nannette Streicher (née Stein; 1769–1833) and her brother Matthäus Andreas Stein, known as André Stein (1776–1842), who had their own separate firms after 1802. After 1823 Nannette Streicher was in partnership with her son Johann Baptist (1796–1871), who continued the business after her death; from 1859 he was in partnership with his son Emil Streicher (1835–1916) who took over in 1871 and dissolved the firm in 1896. Other noteworthy makers included J. J. Könnicke (1756–1811), whose workshop was taken over by Franz Lauterer (*c*1770–1833); Matthias Müller (1769–1844), the number and ingenuity of whose inventions rival those of J. A. Stein in the 18th century; Joseph Brodmann (1771–1848), whose workshop was taken over by his pupil Ignaz Bösendorfer (1796–1859) in 1828 and continued by Ignaz's son Ludwig (1835–1919) from 1859; and Conrad Graf (1782–1851), who in 1804 married the widow of the piano builder Jacob Scheikle and in 1811 moved his workshop to Vienna.

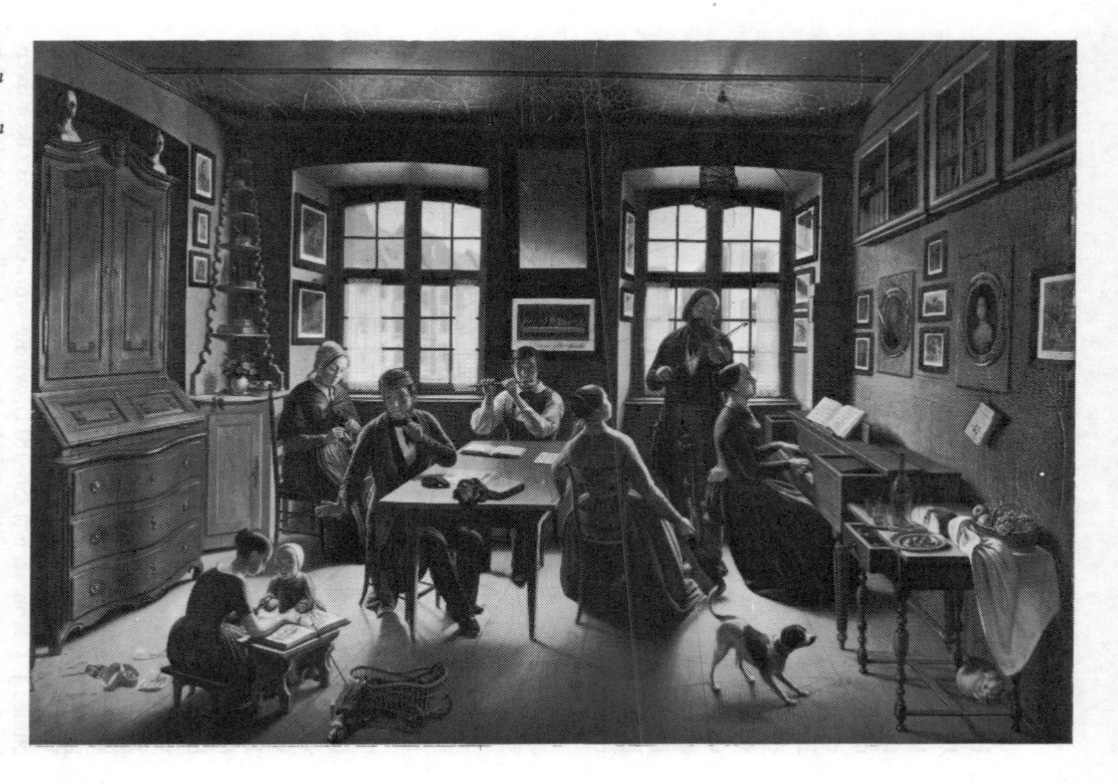

14. Typical mid-19th century south German square piano accompanying a violin and flute: painting, 'Family Concert in Basle' (1849), by Sebastian Gutzwiller

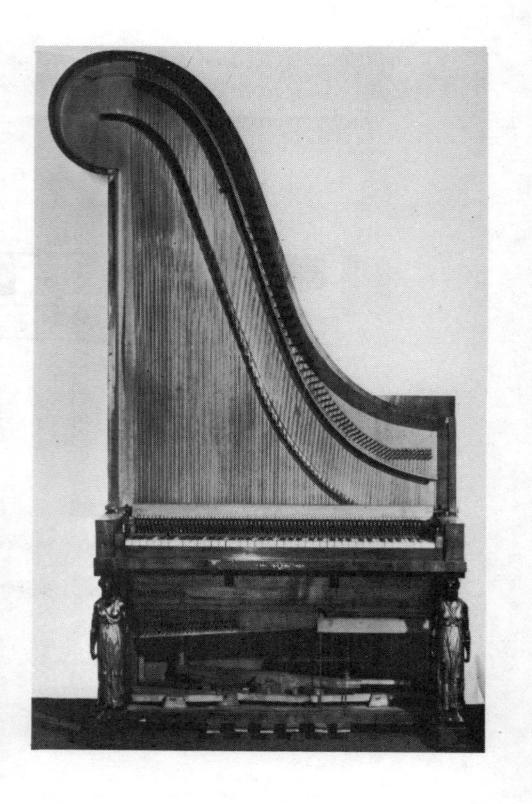

15. Giraffe piano by Ernst Rosenkranz, Dresden, c1815; the right-hand view shows the instrument with the fronts removed to reveal a layout resembling that of a grand piano (here a vertical adaptation of the Viennese action); the pedals (from left to right) are: sustaining pedal, bassoon stop (up to e♭′), full moderator (celeste), half-moderator, una corda, janissary stop

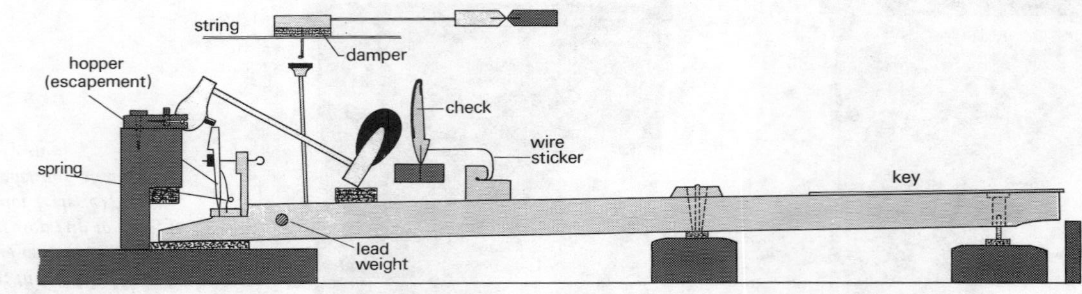

16. Anglo-German action from a piano by Johann Streicher of Vienna, c1845

Several trends of the first half of the 19th century were already discernible by 1800. The five-octave range of the German and Viennese pianos was expanded, and the keyboards were changed from black naturals and white-topped sharps to white naturals and black sharps as on the modern keyboard. The number of tone-altering devices increased. The case structure was made heavier to accommodate the increasing size of the instruments and their heavier stringing.

Few Viennese pianos from the first decade of the 19th century appear to have survived, but several extant instruments by Anton Walter with a range of F' to g''' may be from this period. An early instrument by Nannette Streicher (Germanisches Nationalmuseum, Nuremberg) with a range of five and a half octaves, F' to c'''', has most of the characteristics of a late J. A. Stein piano (see §3 above) including wooden *Kapseln*, but the naturals of the keyboard are ivory, and the grain of the bottom is parallel to the spine. Surviving instruments by Nannette Streicher indicate that about 1805 she adopted the Walter action type with metal *Kapseln* and back checks.

Early examples of signed and dated Viennese action pianos with damper pedals instead of knee levers are those by Nannette Streicher (1814; Germanisches Nationalmuseum, Nuremberg) and Joseph Brodmann (1812; Musikinstrumenten-Museum, Berlin). With one (early) exception the extant pianos by Conrad Graf all have pedals.

By the 1820s a typical Viennese grand piano was nearly 2·5 m long and 1·25 m wide, with a range of six or six and a half octaves and usually with two to six pedals. Certain types of space-saving and decorative upright instruments, such as the 'giraffe' (see fig.15, p.31), 'pyramid' and 'lyre' pianos, were popular, as well as smaller versions of the square such as the *Nachttisch* ('night table'), the *Orphica* (a tiny portable harp-shaped piano) and the *Querflügel* ('cocked hat'). In the second quarter of the century larger squares with the Viennese action were also made, notably by André Stein, who capitalized on the American taste for these instruments.

The 1820s and 1830s were also a time of many inventions and patents for proposed improvements to the piano in Vienna. Soundboard structure, *Kapseln*, the keyboard and down-bearing devices for the nut and bridge seem to have received the most attention. In 1823 J. B. Streicher patented his down-striking action (Pfeiffer's *Zuggetriebe*; see §3 above), of which there are several surviving examples, and in 1831 he invented an 'Anglo-German' action in which the layout of the traditional Viennese action is combined with the action principle of the English piano (see fig.16; this type of action had also appeared in some English and German–Austrian pianos in the late 18th century). Streicher used a system of iron bars in 1835, and Friedrich Hoxa is reputed to have been the first Austrian to use a full iron frame, in 1839; Friedrich Ehrbar (1827–1905) was one of the first at Vienna to use the one-piece cast-iron frame (see §6 below). The influence of these developments on the mainstream of Viennese building, however, would appear to be negligible. Fortunately the Viennese wooden instruments, especially those with interlocking structure, were more capable than the English of sustaining increased string tension. Graf, the most eminent Viennese builder from the early 1820s until his retirement in 1840, remained faithful to wooden framing (see fig.17); two metal braces and a metal hitch-pin plate in one surviving example (Musikmuseet, Stockholm) are later additions. The relative virtues of English- and Viennese-style pianos – their touch and timbre – were keenly and emotionally debated on many occasions. Recent research indicates that composers in Vienna from Beethoven to Schumann and Brahms retained their allegiance to the Viennese piano, as did players well schooled in Viennese keyboard technique. But as the century progressed, the demands of musical taste and the predominant playing technique everywhere else accentuated the supposed disadvantages of the Viennese action. Joseph Fischhof, a juror at the Great Exhibition of 1851, complained bitterly in his *Versuch einer Geschichte des Clavierbaues* (1853) about the other judges' emphasis on volume alone, which discriminated against the already sparsely represented Viennese pianos built to satisfy the Austrian taste for fine nuances and expressive playing.

Just as the demand for more volume with a stronger fundamental tone and fewer overtones meant heavier stringing and consequently a thicker and stronger case structure, the hammers and dampers of the Viennese piano also became heavier (see figs.18 and 19, pp.36–7), although the simplicity of the action did not change and some Viennese makers retained until late in the century a thin layer of leather over the felt hammer-covering common by the middle of the century. Inevitably, however, the heavier action – which could not be counterweighted as with the Erard-type action – diminished that delicacy of touch and crispness of tone which had distinguished the earlier instruments, the mechanism as a whole becoming less responsive. Pfeiffer (1948) suggested that pianists used to the English action were disturbed by the feeling of the hammer falling back to the rest position in the Viennese pianos, which is not noticeable in an action where the hammers are not attached to the key. He also explained that the key-attached hammer had another peculiarity: the striking-point

17. Grand piano (once owned by Robert and Clara Schumann) by Conrad Graf, Vienna, 1839

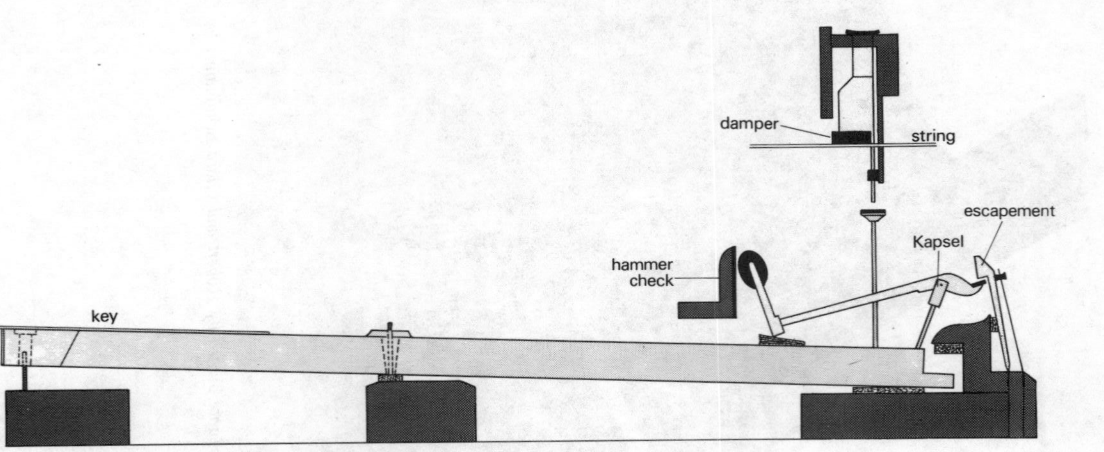

18. Prellmechanik (Viennese action) from a piano by Graf, 1826

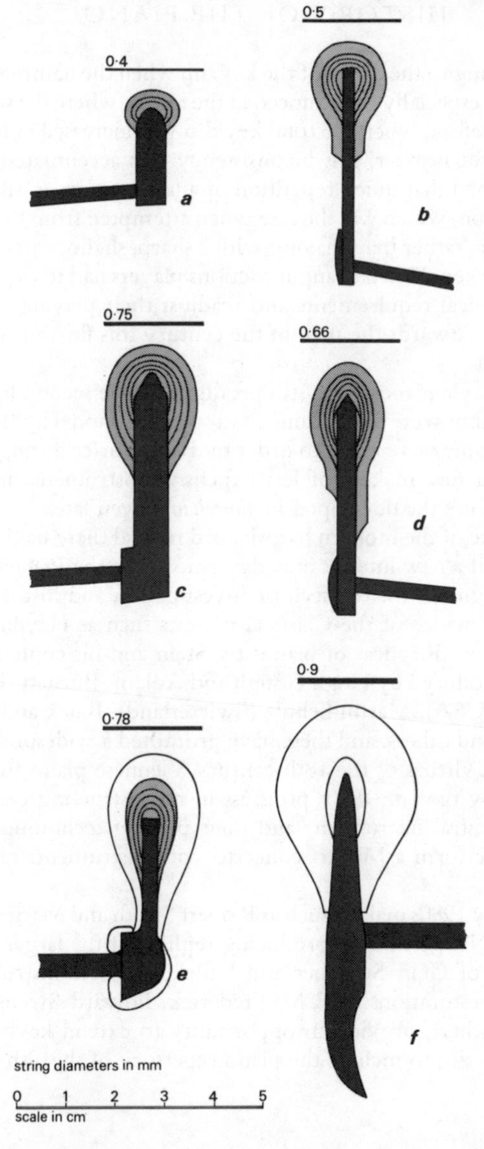

19. Comparison between six piano hammers and strings (all sounding the note f′), showing the gradual increase in mass of both as the instrument developed: (a) south German (Heilman), c1785; (b) English (Broadwood), c1806; (c) Viennese (Graf), 1826; (d) English (Broadwood), c1823; (e) French (Erard), c1825; (f) American (Steinway), c1970

varies according to the depth of the key dip when the hammer hits the string. This is especially pronounced in the treble, where the strings are shorter. Therefore, when the total key dip was increased as the Viennese action got heavier, this inconsistency was accentuated. Players also complained that quick repetition of a note was impossible in the Viennese action, which was the case when attempted from the bottom of the key dip, rather than playing with a sharp, shallow attack on the surface of the key. When changing actions players had to explore each action's technical requirements and readjust their playing technique accordingly. Towards the turn of the century this flexibility became more difficult.

Viennese-style pianos were still produced in the second half of the 19th century but were discontinued as a standard model by Bösendorfer in 1909; some were made to order by Bösendorfer during the next decade and a few makers of less expensive instruments in Vienna continued to use the developed *Prellmechanik* even later.

In the wake of the modern harpsichord revival there has been since World War II a new interest in early pianos with the *Prellmechanik* as proper instruments for the stylistic investigation and historically accurate performance of the Classical masters such as Haydn, Mozart and Beethoven. Replicas of pianos by Stein and his contemporaries have been produced by Hugh Gough and Adlam–Burnett (England), Philip Belt (USA), Martin Scholz (Switzerland), Rück and Neupert (Germany) and others, and these have promoted a widespread recognition of the virtues of the 18th-century Viennese piano for its own repertory. By the late 1970s progress in reconstructing contemporaneous orchestral instruments and their playing techniques made it feasible to perform a Mozart concerto with instruments resembling the originals.

In the early 1980s makers such as Robert Smith and Margaret Hood (USA) and Neupert began producing replicas of the larger Viennese instruments of Graf, Streicher and Dulcken. These instruments, as well as the restorations of E. M. Frederick, Edward Swenson (both USA) and others, provide an opportunity to extend keyboard performing practice to include the piano repertory of the19th century.

6. ENGLAND AND FRANCE, 1800–1860

During the first half of the 19th century English and French piano makers had their greatest period of innovation and growth, laying the foundation for the modern piano before American technology assumed the lead. In Germany, numerous small firms following either Viennese or English principles developed a tradition of craftsmanship which served as the basis for the industry's growth to prominence later in the century, when German instruments came to incorporate 19th-century French and American as well as English inventions.

The introduction of iron bracing around 1820 was crucial for the eventual development of the modern piano with its thick, high-tension strings, large hammers and somewhat bell-like timbre (see fig.19). Long braces, as distinct from mere metal gap spacers between the wrest plank and belly rail as in most late 18th-century English grands, were first developed precisely for the purpose of bolstering the entire instrument's resistance to string tension; but they communicated the stress of that tension to an unduly small area of wood at the point of their attachment. The earliest patent for metal bracing to form something of a complete frame was Joseph Smith's of 1799: metal braces to the case were to be placed under the soundboard without touching it (where the wooden bracing would traditionally have been) to withstand the strain of heavier strings. Broadwood introduced a metal frame in the early 1820s and Pleyel had done so by 1828. Meanwhile John Isaac Hawkins at Philadelphia had suspended the soundboard of his patented upright piano in a primitive form of metal 'compensation' frame (see below). Other early uses of metal included a hitch-pin plate in Broadwood square pianos from about 1808 and an arched metal brace to strengthen the wrest plank; Southwell patented a device of this sort in 1799, and Broadwood used it from about 1815.

A leading article in *The Times* of 7 May 1851 claimed for Erard the invention and first application of metal bars above the soundboard, as distinct from bracing under it, in 1824; but this claim was quickly refuted by Messrs Broadwood in a letter stating that they had applied metal tension bars to the treble of a grand piano in 1808 and had used three to five bars in 1821. Erard obtained a British patent in 1825 for the application of sheet iron and metal bolts, and Broadwood's patent of 1827 added a fourth bar to the three bars or iron stretchers which they had used in combination with a metal string-plate (see fig.20, p.40).

From the 1820s to the 1840s increasing numbers of stress braces were introduced to pianos of all shapes. These braces, of plain rectangular section, were attached to continuous iron hitch-pin plates. Not until the late 1840s was the number of bars reduced, the use of T-section

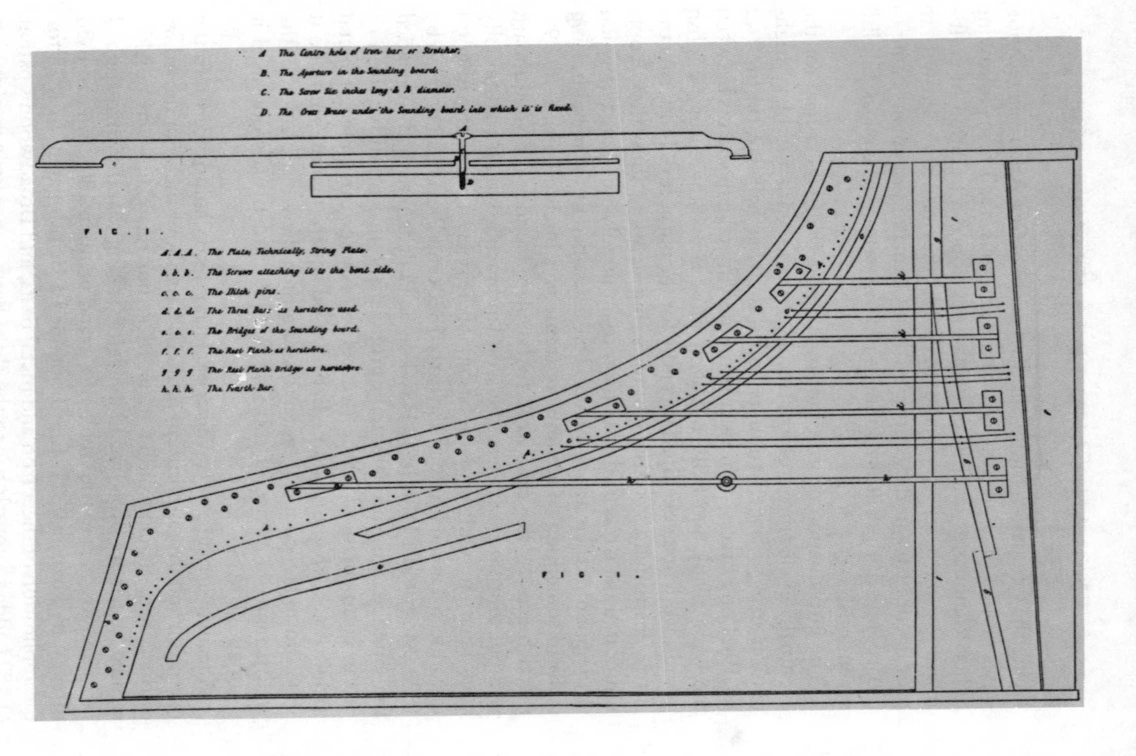

20. *Iron bracing scheme patented by James Shudi Broadwood in 1827*

braces promoted and the bearings between braces and metal hitch-pin plates improved.

The Stodart firm also sent a letter to *The Times* (10 May 1851), referring to their British patent of 1820 for metal 'compensation' frames composed of plates and tubular braces. The object was to reduce the tendency of pianos to go out of tune during changes of humidity and temperature by achieving expansion and contraction of the frame equal to that of the strings. The tension of the brass strings was borne by brass tubes and that of the iron strings by iron tubes; instead of expanding, for instance, with humidity as a wooden frame would, these tubes would, like the strings, expand and contract with changes of temperature and thus prevent the strings from slackening. Stodart applied this system to most types of piano, and Pierre Erard obtained a related French patent in 1822 but built few instruments using the device. Stodart's successful use of it was due mostly to the greater strength of the tubular braces (compared with solid braces of feasible weight) and not to the virtues of the compensation principle itself, which Stodart had abandoned by 1860.

In the quest for greater volume heavier hammers were a natural corollary to the heavier strings which the iron frame could now support. But to prevent a harsh tone these hammers required in turn a deeper cushioning than was provided by the usual layer of deer hide or similar leather. A coarse woollen cloth called molton was experimented with in some English instruments around 1820. Jean Henri Pape obtained a French patent in 1826 for, among other things, felt-covered hammers, and his first subsequent patent included a prescription for preparing the felt. Alternative coverings were tried, including sponge, tinder, india-rubber and gutta-percha; but carefully prepared felt soon became the norm.

To support and exploit the increasingly heavy hammers a stronger and more efficient action was required. Sébastien Erard's repetition action, in which a note could be repeated without the key returning to its position of rest, filled this need. His first repetition action was patented in England in 1808 and a further modification was patented in France the following year; but the action for which he is especially remembered was not invented until 1821, when it was patented by his nephew Pierre Erard in England (see fig.21, p.42). Although this 'double-escapement' action with its elaborate combination of levers was at first not thought to be durable, it is the prototype of all modern actions. The principle is to hold up the hammer at a certain height while the key returns, thus enabling the hopper to re-engage itself under the hammer in less time than in actions with a single escape-

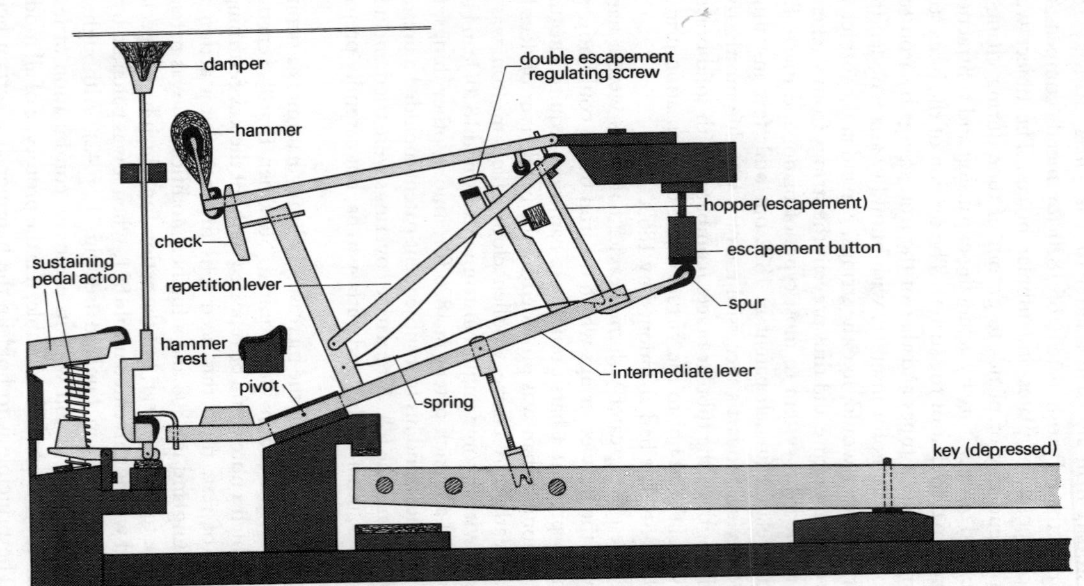

21. Erard double-escapement action after the English patent of 1822; the intermediate lever, pivoted to its flange, simultaneously lifts the hopper and pulls down the damper; the action is shown with the key depressed, the hammer having fallen back to its check

ment. The ways of achieving this were various and other makers invented their own versions; but English makers avoided it until after the middle of the century, and in Pole's description of the pianos exhibited in the 1851 London exhibition those grands that he thought worthy of mention were all German or Belgian copies of Erard's instruments.

In his British patent of 1808 Sébastien Erard included as a detail the invention of the *agraffe*, a separate metal stud for each note, near the edge of the wrest plank, to prevent the up-striking hammers of a grand or square piano from unseating the strings on the nut. A similar purpose is served by the *capo tasto*, invented in 1843 by Antoine-Jean Bord, a fixed rail under which the strings pass just beyond the nut or wrest plank. These devices contributed much to the stability of the instrument.

Much work was also concentrated on down-striking actions (i.e. with the hammers striking from above) to eliminate that structurally troublesome gap between the wrest plank and belly rail through which the hammers in an up-striking action reach the strings. Counterweights or springs were used to keep the hammers poised above the strings when at rest. Sébastien Erard's repetition principle was applied to down-striking actions in patents by Pierre Erard, Pape, Samuel Wirth and others. But eventually upright pianos were the only ones in which it proved feasible to place the hammers on the opposite side of the strings from the soundboard.

At the end of the 18th century the upright piano was basically a grand, set vertically on a stand with the action adapted accordingly. To reduce its height John Isaac Hawkins at Philadelphia and Matthias Müller at Vienna in 1800 independently invented upright pianos with the tuning-pins at the top and the longest strings descending to within a short distance of the ground. Whereas Hawkins, who obtained a British patent as well as two in the USA, adapted the English grand action, Müller used a *Prellmechanik* (he in fact built two types of instrument with perpendicular stringing; the earlier had two keyboards, one tuned an octave higher than the other).

In a British patent of 1802 Thomas Loud suggested 'fixing the first bass strings from the left-hand upper corner to near the right-hand lower corner, and the rest of the strings in a parallel direction' to achieve an upright piano hardly more than 1·5 metres tall. William Southwell patented an improved action for a cabinet piano in 1807. The type of instrument associated with his name is usually about 1·5 metres high with vertical strings to the ground; the hammers are near the top of the instrument, operated from the keyboard by long rods (stickers) in a vertical adaptation of Geib's 1786 action for square pianos. This type of sticker action (see fig.22) was used in the large cabinet pianos popular in England until the 1860s.

In 1811 Robert Wornum (1780–1852), a builder concerned mainly with simple and inexpensive instruments, produced the first upright piano of modern appearance: the 'cottage' piano, just over a metre tall. The hopper worked directly against the hammer, and a projection at the back of the hopper raised the dampers; the strings were placed diagonally, and there were two pedals. This piano was the prototype for many later small uprights including the commercially important French cottage pianos. In 1828 Pape obtained a French patent which included overstringing (Bridgeland & Jardine in New York, 1833, and Gerode & Wolf in London, 1835, were other early proponents of overstringing in uprights). The tape-check action, used by Wornum as early as 1837 and patented by him in 1842, formed the basis of the modern upright action (see fig.23, p.46). Wornum also applied the tape-check principle to down-striking and conventional up-striking grand actions. John Steward in 1841 applied it to his upright piano (a harp-piano known as the Euphonicon) and several French and English makers took it up in the 1840s.

While the compact upright (see fig.25, p.48) quickly became popular among many thousands of lower-middle-class families who could not afford a larger model, it was not yet a satisfactory musical instrument. Square pianos (see fig.24, p.47) grew in size and durability but retained the Geib action. By 1840, however, they were in commercial decline (unlike their iron-framed American counterparts), and none appears to have been made after 1863. A host of curious new models was also produced during the century, including pianos combined with furniture such as tables, beds or writing desks, transposing pianos, pianos with pedal-board, and so on.

In England the grand pianos of Broadwood and Clementi (Collard & Collard after 1830) retained much of their pre-eminence as concert instruments during the first half of the century, and were rivalled in quality, among British makers, by Tomkison in the 1820s and early 1830s. Erard grand pianos, however (see fig.26, p.49), challenged them in England – there was an Erard factory in London as well as Paris – and quite superseded them in France. A certain vein of fluent Paris salon virtuosity, exemplified by such pianist–composers as Kalkbrenner and Thalberg, exploited the smoothness of touch afforded by Erard's double-escapement action (for which the British patent of 1821 was renewed in 1835). The harmonic and textural elegance of Chopin's music was also particularly well served by the suavity of the

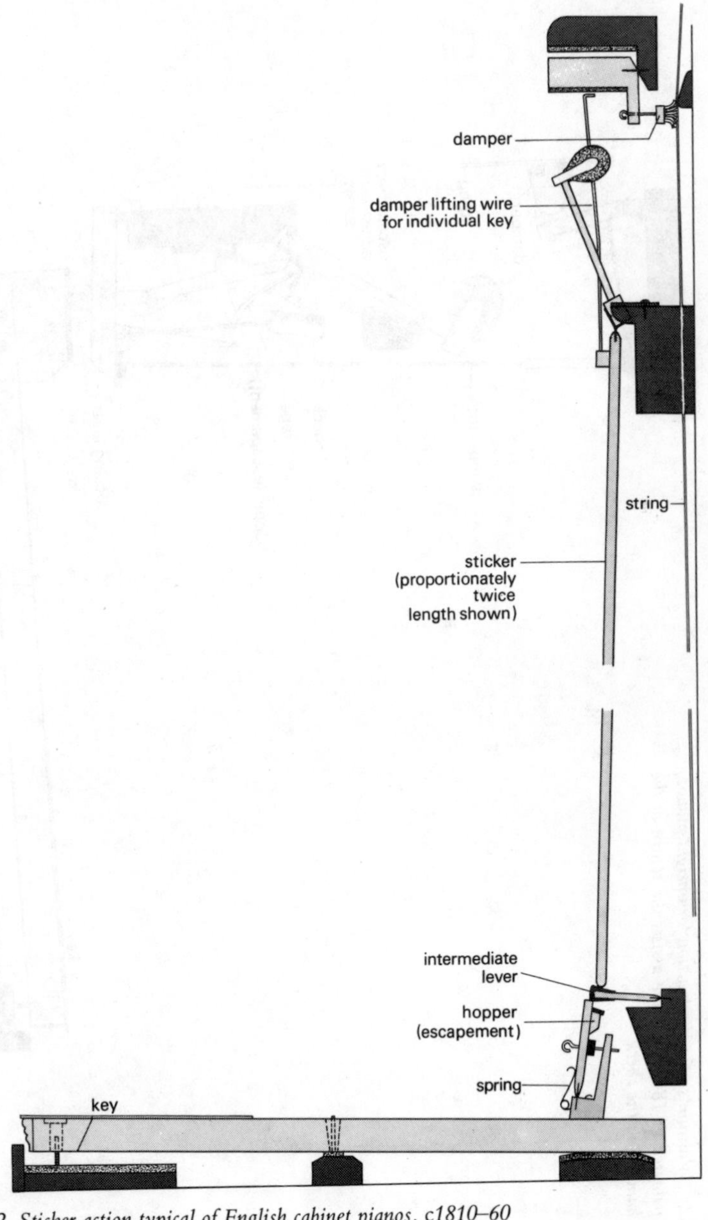

22. Sticker action typical of English cabinet pianos, c1810–60

23. Wornum's tape-check action for upright pianos, patented in 1842; the tape assists the return of the hammer to the check

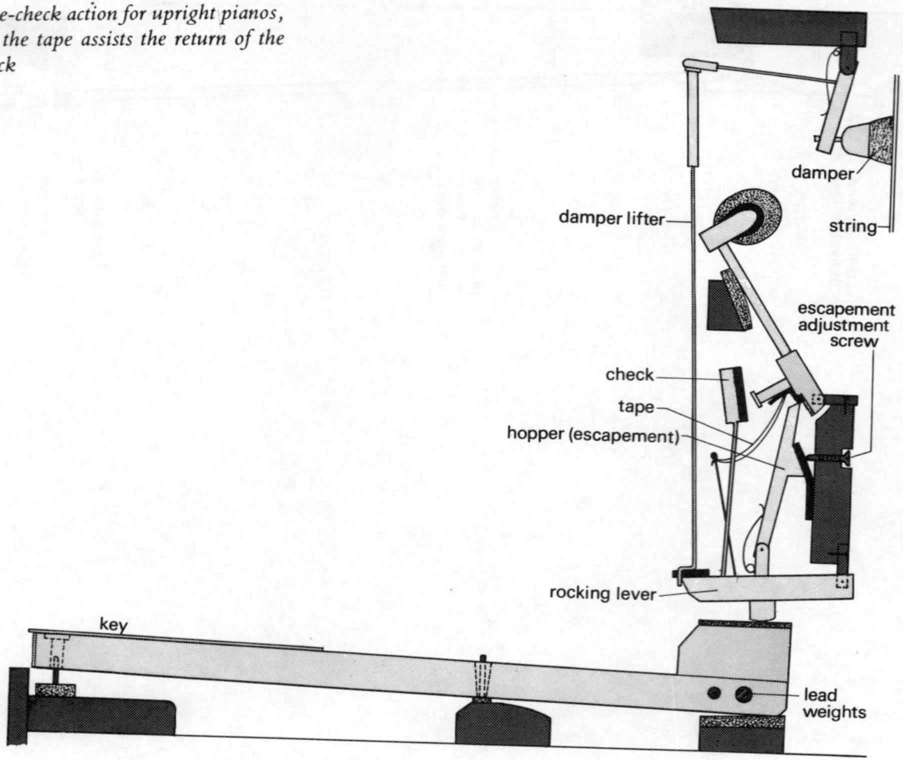

24. Square piano by Clementi & Co., London, after 1822

25. Cottage piano by Clementi & Co., London, c1825

26. Grand piano by Erard, London, 1855

best French pianos of his period, though Chopin usually preferred the instruments of Pleyel, who had begun making pianos in 1807, to those of Erard.

7. NORTH AMERICA TO 1860

The earliest known reference to a piano in North America is in the *Massachusetts Gazette* of 7 March 1771. Later that spring, Thomas Jefferson asked Thomas Adams in England to purchase on his behalf a piano instead of the clavichord he had originally ordered for his fiancée. A piano was used at a concert in New York in 1773, and according to a note in the *New York Journal* (1774) a 'set of hammer harpsichords slightly damaged' was included in a list of goods from the wrecked ship *Pedro* that were sold at auction. John Jacob Astor imported music and musical instruments including pianos to New York from the late 1780s for a few years until his fur trade increased, and there were other part-time dealers.

In 1775 Johann Behrent, a German immigrant living in Philadelphia, built the first American piano, a square; but Charles Albrecht was the first important maker, working in Philadelphia from about 1790 and making close copies of English instruments. Charleston had a piano maker as early as 1791. In 1792 Dodds & Claus of New York advertised a piano that could withstand the rigours of the American climate, a challenge that later became one of the driving forces of the industry; and an American patent for certain (unknown) 'improvements to pianofortes' was granted in 1796 to James Sylvanus McLean. John Isaac Hawkins, who worked in Philadelphia around the turn of the century, invented an upright piano and a compensation frame (1800; see §6 above), both of which he patented in England as well as in the USA.

Boston supported an active trade during the first two decades of the 19th century and became the most important centre for the development of the American piano to the 1850s. The Franklin Music Warehouse, established about 1813 by Thomas Appleton, Alpheus Babcock and a certain Hayts to make pianos as well as import English instruments and printed music, began to produce cabinet pianos in larger quantity than anywhere else outside Britain. This part of the enterprise was not commercially successful, however, and soon the firm's piano makers were obliged to seek work elsewhere. Babcock himself worked for a while for John Mackay, a wealthy sea captain interested in pianos. In 1824 Babcock won first prize for an instrument displayed in Philadelphia at the Franklin Institute's first exhibition, and in 1825 he patented the cast-iron frame, with hitch-pin plate, for a square piano. New York manufacturers were, like their European counterparts, reluctant to take up this invention, but during the 1830s it was adopted by Boston makers including Jonas Chickering, who in the late 1820s had joined in partnership with Babcock's former partner, Mackay. With Mackay's commercially shrewd help Chickering built up the most successful American firm until Steinway. He patented his own iron frame for square pianos in 1840 and then turned his attention to the grand piano, an instrument previously neglected in the USA. He received a patent for a full one-piece iron frame for grands in 1843 and for a decade he remained the only significant American producer of grand pianos. William Knabe, another German immigrant, established a firm in Baltimore in 1839 which, although not innovative, produced instruments of high quality incorporating the latest technical advances. Knabe pianos became particularly important in the musical life of the southern states by the outbreak of the American Civil War.

The large square pianos were the first distinctively American contribution to the history of the instrument. Having first been equipped with Erard's repetition action by 1830, and felt-covered hammers soon afterwards, they developed, thanks to their iron frame, into large, heavy and fairly sonorous instruments (see fig.27, p.52) – a far cry from their clavichord-like, 18th-century German predecessors. They were not quite as resilient as their weight might suggest, however, and their systems of overstringing rather crowded the strings, or sometimes even employed two soundboards.

In 1853 Heinrich Steinweg, a German piano maker obliged to emigrate in 1850 because of political unrest, established the firm of Steinway & Sons at New York. The firm expanded rapidly, built grand pianos, and won a gold medal for overstrung square pianos at the 1855 American Institute Fair. In 1859 the younger Henry Steinway obtained a patent for an overstrung grand piano (see figs.28 and 29, pp.53 and 54–5). The same year he wrote to his brother Theodore (who remained in Germany until the death of Henry and Charles Steinway in 1865): 'Our overstrung grands are really excellent.... You will soon see a rave article about us and our business in the *Berliner Musikzeitung* and the *Leipziger Signale*'. Steinway's overstringing, in which the strings were crossed in a fan-like pattern, used a conventional soundboard but shifted the bridges nearer to its centre to avoid a clutter of strings and make better use of the capacity of the soundboard

27. Overstrung square piano by Steinway & Sons, New York, 1877–8

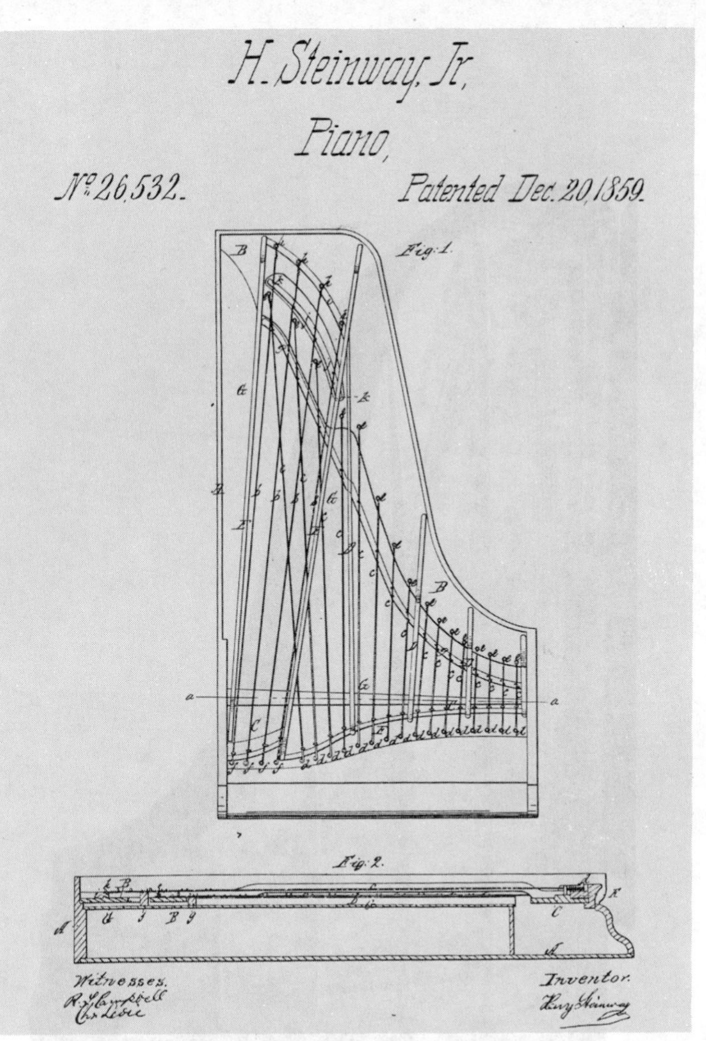

28. Detail from Henry Steinway's patent for an overstrung grand piano, dated 20 December 1859

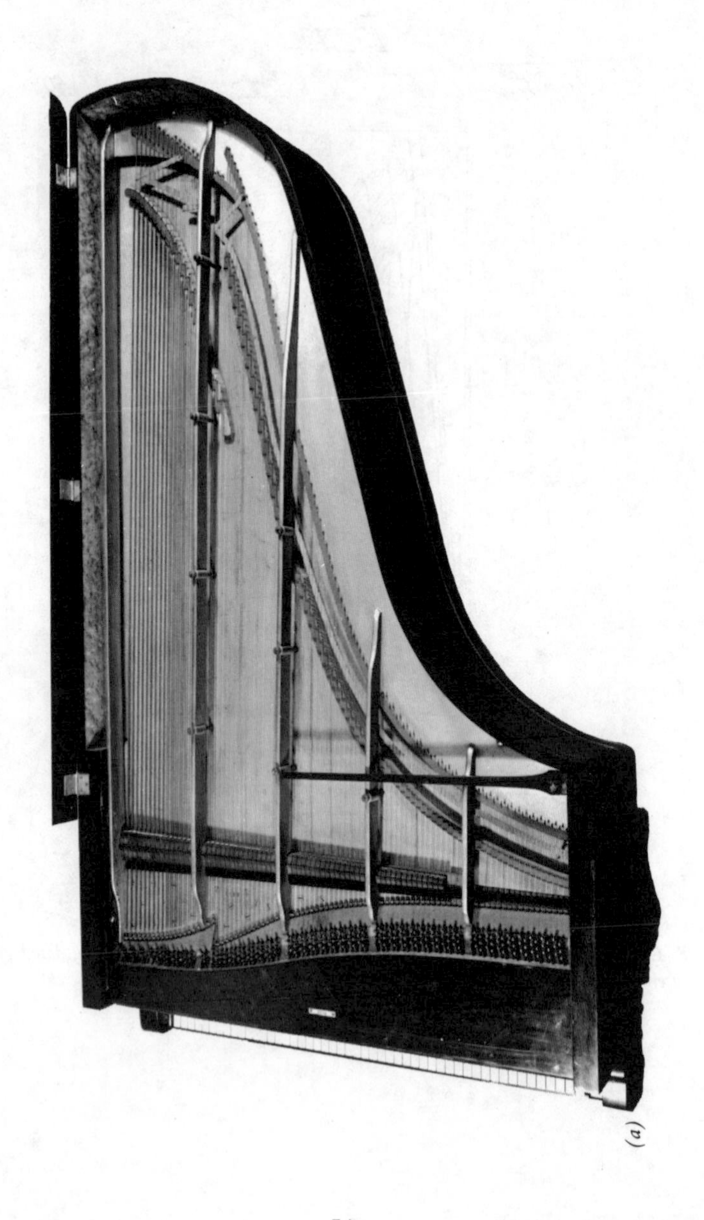

(a)

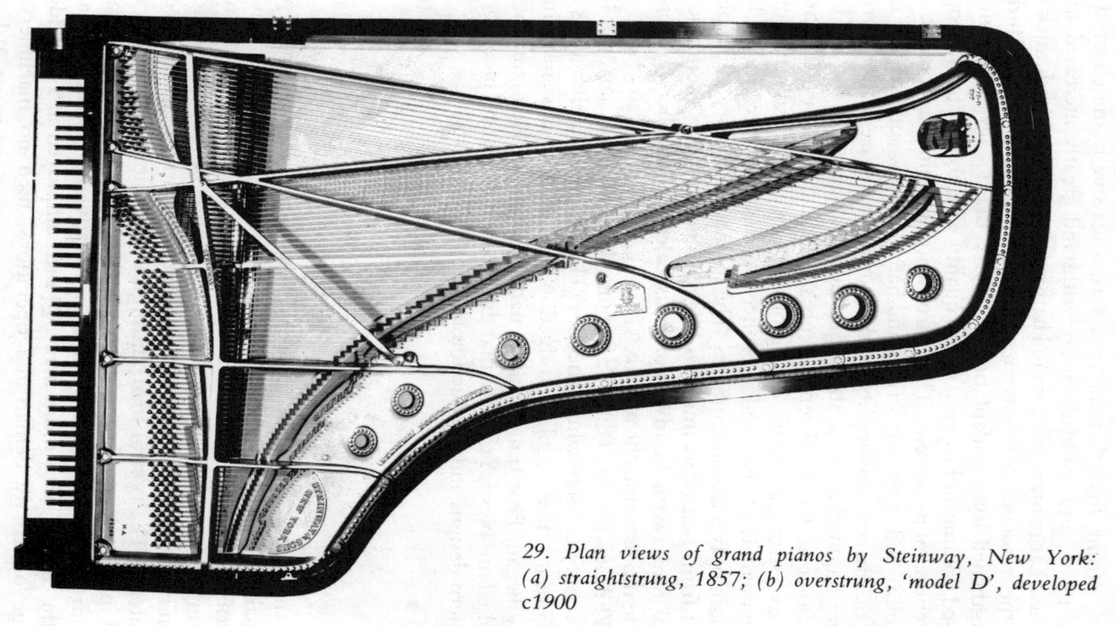

(b)

29. Plan views of grand pianos by Steinway, New York: (a) straightstrung, 1857; (b) overstrung, 'model D', developed c1900

to enhance the instrument's sound. The new grands were more heavily strung than their predecessors; a massive frame covered the wrest plank, and could therefore withstand greatly increased string tensions. Large machine-covered hammers were fitted with thick felt, and the resistance of heavier strings to their impact was further strengthened by an adaptation of Pierre Erard's agraffe. Their increased volume and ringing sound quality were to transform people's conception of piano tone. The instruments also gained a reputation for reliability and resilience; strings and hammers now survived the most athletic virtuoso onslaughts, so that tuners and spare pianos were no longer required to be in attendance during concerts.

Numerous technical refinements since the invention of the Steinway overstrung grand have promoted stability and durability, but the fundamentals of the modern concert grand were present by 1860, albeit in only a few instruments. The subsequent history of the piano, apart from that of its music, is largely a matter of its economic and social impact with the advent of industrialization and the development of the modern upright models.

While it is recorded that the first piano was made in Canada in 1832, it was only in 1860, when Theodore Heintzmann established his firm in Toronto, that regular production there of first-class instruments truly began. First trained in his native Germany, Heintzmann later worked in New York so that his Canadian pianos incorporated modern designs and production technology.

8. 1860–1915

In the period from 1860 to World War I a new technology of piano making arose to support a great extension of production and distribution and a fundamental change in the 'balance of power' among the principal countries of manufacture. In 1850 pianos had still been luxury goods, made by craftsmen for musicians and prosperous amateurs. The typical firm, distinguished or nondescript, was small and inefficient, employing labour-intensive methods to make a few hundred instruments each year. Even Broadwood's factory, the largest in the world, with more than 300 men employed in a bewildering range of strictly demarcated operations, used no machinery and achieved virtually no economies of large-scale production. Wherever pianos were made or sold, costs and profit margins were high, turnover low, and prices therefore extravagant: £140 for a grand, more

TABLE 1: Estimates of piano production, 1870–1980 (in thousands)

Year	*Britain*	*France*	*Germany*		*USA*	*Japan*	*USSR*	*Korea*
*c*1870	25	21	15		24			
*c*1890	50	20	70		72			
*c*1910	75	25	120		370		10	
*c*1930	50	20	20		120	2		
*c*1935	55	20	4		61	4		
			W	E				
*c*1960	19	2	16	10	160	48	88	
*c*1970	17	1	24	21	220	273	200	6
*c*1980	16	3	31	28	248	392	166	81

Note: estimates for 1960, 1970 and 1980 are taken from United Nations, 'Growth of World Industry' and 'Yearbook of Industrial Statistics'; for earlier years see Ehrlich, 1976

than £50 for a tolerable upright. A clerk or skilled artisan would need to save practically a year's income to secure an instrument. Some were cheaper, but they were inevitably shoddy and inadequate. Several of these were shown at the 1851 exhibition, including a makeshift 'artisan piano' with reduced compass and uncovered keys, suggesting that a demand existed but could not yet be satisfied. Total world output was probably fewer than 50,000 instruments a year. English, French and Viennese manufacturers were the acknowledged leaders; American and German makers were few and not yet highly esteemed. By 1914, as Table 1 indicates, a drastic transformation had taken place. In this golden age for the consumer, most pianos were factory products that many people could afford. The London price of a good German grand was below £100. A superb Bechstein 'model 9' upright (see fig.30, p.58) cost approximately £50; excellent overstrung instruments were available for £30 and serviceable pianos could be procured at half that figure. Comparison with the general level of incomes indicates that the real cost of pianos had approximately halved since 1850. Moreover, these instruments were more durable: many survive to the present day.

Improvements in the quality of upright pianos were particularly striking. Theodore Steinway caused the firm to introduce fan-like overstringing into an upright model in 1863; the design was improved and output increased during the early 1870s; and in 1882 an advertisement justly claimed that their uprights compared with predecessors 'as a modern ironclad to a canoe'. The application of the new technology to uprights had profound economic implications because it became an essential condition of business success for any piano-making firm and because its basic elements were capable of rapid diffusion throughout

30. Upright piano ('model 9') by Bechstein, Berlin, manufactured during the 1930s

the industry so that instruments by the best makers were no longer necessarily separated by an immeasurable gulf from those in general use.

Several American piano manufacturers, including Chickering, had appeared at the 1851 exhibition, and Steinway took a prize in London in 1862. But it was at the Paris exhibition of 1867 that the ascendancy of what contemporaries described as 'the American system' was fully acknowledged. Of four gold medals awarded by a predominantly conservative jury, two were for American pianos, one for a Viennese copy of a Steinway, and one for a traditional English instrument, described in a prophetic phrase as a 'souvenir des travaux passés'. This victory for innovation was consolidated at the 1873 Vienna exhibition, not directly by American manufacturers, who were poorly represented, but by a host of German instruments modelled upon those of Steinway. A decade later the American system's advantages, which had again been overwhelmingly demonstrated at the Amsterdam exhibition of 1883, were ably summarized in Victor Mahillon's report: quality was maintained and costs lowered by economizing in the use of expensive skills, substituting machinery for manual work, and by a thorough-going division of labour.

These processes were extended on an international scale through a marketing system that was so well articulated that even small makers could gain access to many of its benefits. An almost unlimited range of 'supplies' became available on credit and in any desired size and quantity, listed in clear multi-lingual catalogues: iron frames, soundboards, seasoned timber which had been cut and shaped, veneers, and myriad small parts. Of decisive importance was the highly competitive supply of complete actions, in which several German and French firms excelled. Previous generations of piano makers had been forced to assemble these intricate mechanisms individually and at great cost, or else to skimp (with disastrous effect on the quality of the product). Now for a few pounds they could purchase actions 'ready made' but tailored to their specific requirements. This ready flow of supplies was used throughout the industry, and it enabled small makers to assemble pianos with little capital outlay and a limited range of skills. If the results were not equal to those coming from the great houses, they were nonetheless distinctly superior to most of their laboriously manufactured predecessors.

There were considerable divergences in national response to the new technology. In the USA and Germany it was unequivocally accepted by new makers, who were unhindered by traditional practice or self-esteem. The leading firms acted as pacemakers of technological advance and spearheads of trade, creating an image of modernity and musical quality which benefited their compatriots, some of whom openly claimed allegiance: pianos were advertised 'after' the Steinway system. Bechstein, Blüthner and Steinway, all founded in 1853, were dominating concert platforms and the high-quality market by the 1870s, while continuing to innovate and improve.

The growth of American production – by 1900 more than half of the world's pianos were made in the USA and the five largest manufacturers were all American – was part of a general economic prosperity, with a buoyant domestic market protected from foreign competition by high tariffs and extremes of climate and domestic heating which ravaged the few European instruments to be imported. Many excellent makers, such as Knabe and Mason & Hamlin, never attempted to establish reputations in Europe. The latter firm, established in 1854 as a maker of reed organs, introduced upright pianos in the early 1880s and soon thereafter began producing grands, ranking as concert instruments with those of Steinway, with exceptional success. In 1891 Baldwin in Cincinnati began to produce inexpensive upright pianos, but soon also made grands of such high quality that they were awarded the Grand Prix at the 1900 Paris Exposition. Technical innovations were introduced continually during the period by a number of American makers: some were successful, notably the Gertz soundboard tension resonator system patented by Mason & Hamlin in 1900, but others, like the same firm's screw tuning-pins (which harked back to Taskin's similar invention, previously imitated by Pleyel, Brinsmead and others) were abandoned.

A unique feature of the American market was its loyalty to square pianos (many of them fine large instruments) throughout the third quarter of the century. In the late 1860s over 90% of all pianos produced were squares; but by the 1890s their proportion was negligible. The far smaller Canadian market was served by a number of domestic makers, active largely in Toronto. Instruments of high quality continued to be produced by Heintzmann as well as by younger firms, such as Mason & Risch, founded in 1871.

Apart from a greater dependence on foreign trades, German development followed the American pattern: while exports flourished, the industry's rapid growth was predominantly based on a prosperous home market which stimulated makers to adopt the new technology. They enjoyed the unique benefits of a national system of technical and commercial education and widespread respect for applied science and music. The great prestige of German music naturally attached itself to the country's pianos and the image was a faithful reflection of reality. In addition to the recognized leaders – Bechstein, Blüthner, and the Hamburg branch of Steinway – there were several makers of comparable excellence, including Feurich, Ibach, Grotrian–Steinweg and Lipp, and many others who also built high-quality pianos. No German firm achieved the output of the largest American manufacturers, but at least a dozen were producing more than 1000 instruments a year by 1900, confounding the common myth that high quality could be achieved only by small masters.

In France the course of events could hardly have been in greater contrast. Innate conservatism and a national predilection for refined 'thin' tone (comparable perhaps to French styles of singing) encouraged the leading makers to adopt an attitude of lofty indifference or disdain towards the new technology, reinforced by chauvinism and the constraints of a protected but narrow and stagnating home market. Sales of French pianos abroad collapsed irrevocably. Erard, forced to close its London factory in 1890, barely increased its total production during a period when world markets were rapidly expanding: an extraordinary record for a business that in 1850 was one of the greatest in Europe and had once been at the forefront of technical advance. Pleyel, probably the best French maker of modern times, was less conservative but still unable to sell many instruments outside France. Bord, once famous for sturdy, cheap pianettes, attempted to diversify production without much distinction. Gaveau was ambitious and commercially successful at home, but failed to win acceptance by foreign musicians.

In England the response to the new technology of established manufacturers, guardians of an obsolescent tradition, was similarly unwelcoming. As in France, the leading firms were the least adventurous. Some, like Kirkman (formerly Kirckman) in 1896, were driven out of business; others, like Collard, experienced dwindling sales with varying degrees of equanimity. Expensive casework (as in the Alma Tadema Broadwood; see fig.31, p.62) and excellent workmanship failed to compensate for obsolete technology. Exports suffered, even in the well disposed, prosperous Australian market. Entrepreneurial failure was conspicuous because the English market, though more open to foreign competition than the French, was also much larger and more buoyant. A rapid increase in purchasing power was enhanced by the growth of hire purchase, an important social revolution pioneered by sewing machines and pianos. In the absence of leadership, response to these opportunities had to come from humbler makers. By the turn of the century a 'new school' of English manufacturers had appeared, overcoming deficiencies of capital,

31. Engraving (1879) by Lawrence Alma Tadema of his own Broadwood piano in 'Byzantine' style, inspired by the Basilica of St Sophia, Istanbul

skilled labour and appropriate cultural background. The success of such makers as Hopkinson and Rogers, at home and abroad, was primarily in the 'medium class' market, but a few were more ambitious, making high-quality uprights (Marshall & Rose and Chappell are representative), small grands, and even concert instruments. Half the pianists appearing during the 1910 Promenade concert season played pianos by Chappell, whose uprights were also equal to all but the best German instruments.

9. FROM 1915

It is a commonplace that pianos lost much of their social prestige as domestic instruments some time between the world wars. An explanation would have to disentangle the effects of war, economic crisis and the bewildering array of new goods and services which, competing for the attention of consumers, usurped the piano's sovereign status. Briefly it can be said that the demand for them was eroded by alternative forms of entertainment and 'conspicuous consumption', and by the Depression of the 1930s.

During World War I and the immediate postwar years, full employment in England generated a brisk demand, but manufacturers had become so dependent on imported components that protection from German competition was of little benefit until the components could be produced at home. Meanwhile supplies faltered, prices rose, and quality fell to unprecedented depths. After the war an efficient action-making industry was at last established, and a few firms again began to make reliable instruments, aided by their experience of modern industrial procedures in the production of aircraft. In France output soon regained its modest pre-war level of some 20,000 instruments a year. Inevitably it was a defeated Germany that experienced the greatest turbulence, yet by 1927 annual production had again passed the 100,000 level and leading makers had regained their former pre-eminence. By European standards the American industry was comparatively unaffected by the war, though the best makers were forced to restrict production rather than relax standards. As in Europe, but to a far greater degree, the 1920s were the last great age of piano ownership. Production reached a postwar peak of 347,000 in 1923; a few years later it was estimated that half the country's city dwellers owned an instrument. Over-production, always a risk with durable consumer goods, was temporarily abated by an astonishing rise in the

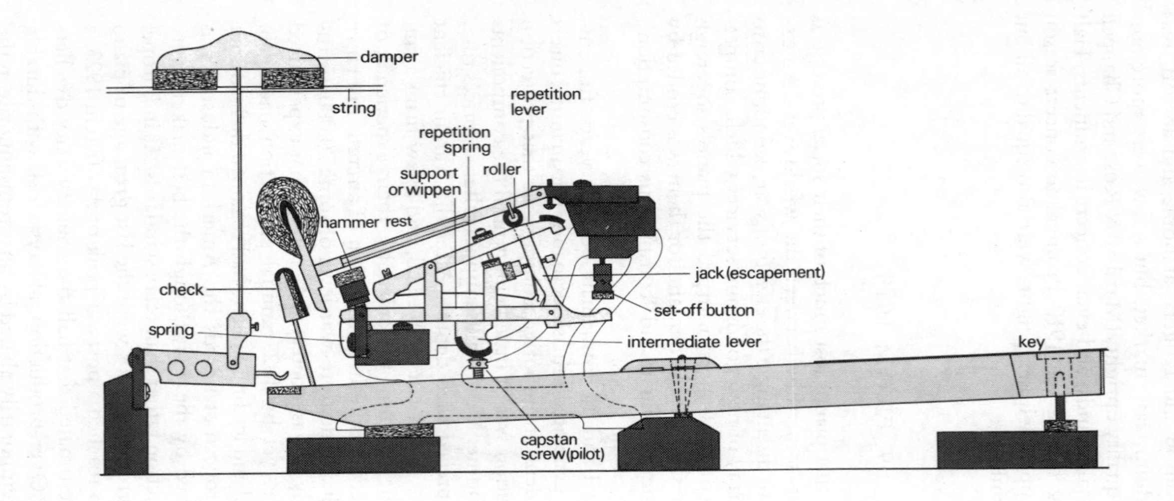

32. Action of a modern grand piano: on pressing the key the movement is transmitted via the pilot to the intermediate lever; the jack then acts on the roller of the hammer which rises towards the string. The moment the backward projection of the jack contacts the set-off button the jack moves back permitting the hammer to escape and to continue in free flight to strike the string and then begin its descent; it is then caught and retained by the check and repetition lever as long as the key remains depressed. If the key is partly released the hammer is freed from the check, and the roller is acted on directly by the repetition lever; it is thus possible to strike the string again by depressing the key a second time (the jack will re-engage with the roller only when the key has been fully released so that a full hammer stroke may be made).

33. Action of a modern upright piano: the movement of the key is transmitted directly to the jack, which in rising pushes the hammer forward towards the string until the backward projection of the jack reaches the set-off screw, thus allowing the jack to escape and the hammer to fall back from the string to be caught by the check; the return of the hammer is assisted by the tape which is so adjusted to jerk the hammer away from the string at the moment of impact

34. Baby grand piano (the 'Elfin') by Broadwood, London, manufactured in 1924–30

popularity of automatic pianos: during the early 1920s sales actually exceeded those of conventional instruments. Despite considerable inventiveness and interest in Europe the craze for player pianos was very much an American phenomenon, which collapsed before the end of the decade.

A more fundamental reorientation in patterns of consumption was also first perceived in the USA where, even before the war, motor cars were recognized by the piano trade as formidable competitors for status. Within a generation they became first a middle-class obsession, and then the supreme object of aspiration for the common man. Gramophones were more obvious rivals, for entertainment, if not status. As techniques and repertory improved they ceased to be regarded as toys and began to challenge the piano as a cheaper and less demanding household orchestra, stimulating musical appreciation but deterring the amateur. Similar intrusions on expenditure and time, on a vastly greater scale, came from the cinema and radio. The process of displacement, which was gathering speed by 1929, was accelerated by the ensuing Crash and Depression. Between 1927 and 1932 American production of pianos fell from 250,000 to 25,000; in Germany the number of manufacturers fell from 127 to 37, and output from 100,000 to 6000. The English decline was only comparatively less steep, from 92,000 to about 30,000.

Among surviving manufacturers it was generally conceded that the demand for pianos would henceforth be based primarily on musical needs, in concert halls, studios, schools, hotels and ships. But there were also some hopes of a fashionable market, for which a new image was required. Baby grands, all less than 1·6 metres in length and some absurdly foreshortened (see fig.34), achieved a measure of success among people more concerned with appearance than with intrinsic quality. More significant, commercially and for its long-term influence on design, was a rash of miniature uprights which began in 1934. Early models were retrogressive in every respect except superficial appearance. It was rational to reduce size, weight and power to fit modern living conditions, but in practice the first miniature pianos failed to meet even modest musical requirements. Reducing their height below a metre necessitated a 'drop' action which gave a 'spongy' unresponsive touch. Tinny in sound, the instruments were also difficult to tune and repair, but these limitations did not hinder the successful exploitation of a new market which employed clever advertising to emphasize the product's modernity. Several foreign makers followed suit, and even the USA adopted this English innovation, thus for the first time in a century reversing the direction of technological migration. External design was different from European patterns, and the American miniatures were called 'spinets', but the essential features, limitations and success were the same. By the late 1930s they dominated markets everywhere.

Good pianos were still being made, however. Indeed it is arguable that this was the last vintage period when high skills and fine materials were still abundant. Two firms deserve special mention, though they represent quite different levels of achievement. Bösendorfer (see fig.35, p.69), a familiar name in Vienna since 1828, became the first Austrian maker to attain international prominence in modern times. In 1936 they won first place in two categories of a BBC competition to select pianos for broadcasting, and their instruments are still characterized by artistic excellence. Rather less exalted but nonetheless propitious in its own way was the arrival of Alfred Knight, a British maker whose small uprights used techniques and materials that had been developed during World War II – plastics, laminates and new adhesives – and began to establish a reputation for good workmanship.

The most important development of postwar years, however, came from Japan. In 1953 10,000 pianos were produced there. Within a decade output increased tenfold, and in 1969 Japan became the world's largest manufacturer, with a total of 257,000 instruments, some 35,000 more than the USA. Japanese pianos have a longer history than is generally recognized in Europe, a significant fact in an industry that places an irrationally high premium on a firm's antecedents. The two leading makers, Yamaha and Kawai, were founded in 1887 and 1925 respectively, while Japanese interest in Western music, as well as in more material aspects of the West, dates back to the Meiji restoration of 1868. The government's educational policy was of particular importance: the training of teachers included lessons in the piano, organ and violin. A Japanese square piano was shown at the 1878 Paris exhibition, and by the 1890s large numbers of reed organs were being manufactured, piano actions imported, and apprentices sent to study in the USA. By 1910 Yamaha were producing several hundred pianos, including a few small grands 'following the Steinway model', some for export; and foreign instruments had been entirely replaced in schools. The essential features of Japanese piano manufacture had therefore appeared long before the expansion of the 1960s. The home market was predominant – even the huge exports of the mid-1970s accounted for little more than 10% of output – and a firm link with education was retained, manifested in the 6000 Yamaha 'music schoolrooms'. There was much copying, particularly of American models, but there was also a rapid emancipation from foreign components and dependence. Concert instruments were developed more slowly, but some were made before 1914; Japanese concert grands have now been played by some of the world's leading pianists. By the mid-1970s Yamaha's annual output of some 200,000 instruments was far greater than even the largest American manufacturers' at the height of their activities.

By 1970 world production was actually greater than in the peak years before 1914, yet the market for pianos had not grown commensurately with the growth of population and incomes, nor with the demand for other products. In Western countries there is no sign of a resurgence of the piano mania that characterized the early 20th century and the demand for instruments is not affected by a desire for status.

35. Grand piano ('model 290, Imperial') by Bösendorfer, Vienna, first manufactured c1900

10. DEVELOPMENTS AND MODIFICATIONS

Musical needs have resulted in a number of experiments carried out with the conventional piano for different purposes. An early modification, which survives today in pianos manufactured by Steinway and many American companies, is the third or 'sostenuto' pedal, first introduced in 1844 by Boisselot in Marseilles, but not established until the American branch of Steinway adopted it in 1874. Bösendorfer grand pianos have an extension of a minor 6th to the normal range in the bass for reinforcing the left hand with octaves; a hinged flap covers these keys when they are not required, to prevent confusion in the player's visual orientation.

Several modifications to the shape and layout of the keyboard, intended to simplify fingering, were tried out in the 19th century. The 'sequential keyboard', invented by William A. B. Lunn under the name of Arthur Wallbridge in 1843, aimed at reducing the supremacy of the C major scale. Each octave included six lower keys, for C♯, D♯, F, G, A and B, and six raised ones, for C, D, E, F♯, G♯ and A♯. A similar arrangement was advocated by the Chroma-Verein des Gleichstufigen Tonsystems in 1875–7. Paul von Janko's keyboard (1887–8) is a later application of the same principle. The two rows of keys were triplicated, providing a total of six rows, each slightly higher than the other and each including six keys in the octave. This arrangement permitted the same fingering in all tonalities. Jozef Wieniawski designed a piano with reversed keyboards, patented by E.J. Mangeot in 1876, which was actually made of two superposed pianos, one with the treble at the right as usual and the other with the treble at the left. The purpose was to permit the same fingering for the same passages in both hands. This arrangement is reminiscent of some medieval representations of keyboard instruments where, for reasons that remain unclear, the treble is shown at the left. In 1907 F. Clutsam patented a keyboard with keys arranged in the shape of a fan according to a principle already conceived by Staufer and Heidinger in 1824 and supposed to facilitate playing in the extreme bass and treble.

Although the first microtonal piano was constructed in 1892, it was not until the 1920s that a considerable number of quarter-tone pianos were built (see Table 2), as well as a few in other tunings. Some of these instruments have two manuals tuned a quarter-tone apart; others have three manuals, the third duplicating the first to allow alternative fingerings (the length of the keys diminishes on each manual, so that on the one furthest from the player the white keys are the same size as the black ones). As with the two manuals of the Emanuel Moór piano (of *c*1920), which are tuned one octave apart (with provision for coupling them together) and simplify the execution of octaves and other large intervals, these permit each hand to play notes on two manuals simultaneously.

A series of *pianos metamorfoseadores* (microtonal upright pianos with conventional keyboards), each in a different tuning from $\frac{1}{3}$- to $\frac{1}{16}$-tones, was planned by Julián Carrillo in 1927; a $\frac{1}{3}$-tone grand was built in 1947 and the uprights (by the Carl Sauter Pianofortefabrik in Spaichingen, Baden-Württemberg) in 1957–8. The range of these pianos becomes smaller as the number of subdivisions of the octave is increased, so that the $\frac{1}{16}$-tone instrument has a compass of a single octave, in the middle range, with 97 keys. Many of Carrillo's instruments are to be housed in the Carrillo Museum in Mexico City. Between the 1930s and 1960 Augusto Novaro, a former pupil of Carrillo's, built a number of Novares – pianos that sound less percussive than normal and are tuned in such divisions of the octave as 14, 15, 19, 22, 31 and 53. Most contemporary composers and performers prefer to use conventional instruments that are retuned or specially fingered. Ben Johnston's Sonata (1963) requires a piano tuned in a just system in which only seven pairs of keys, mostly several octaves apart, give octave relationships. Just intonation is also used in La Monte

TABLE 2: Quarter-tone pianos, 1892–1931

No. of manuals	*Date*	*Inventor, builder*
2 manuals 'achromatisches Klavier'	1892	G. A. Behrens-Senegalden (Berlin)
(unfinished)	*c*1913–14	Arthur Lourié (St Petersburg), built by Maison Diederichs
2 manuals (unfinished)	1922	Ivan Vïshnegradsky (Paris), built by Pleyel
2 manuals	1924	Moritz Stoehr (New York)
3 manuals	1924	Alois Hába (Berlin), Ivan Vïshnegradsky (Prague), built by Grotrian-Steinweg
2 manuals	1924	Alois Hába (Prague), built by Förster
3 manuals	1925	Alois Hába (Prague), built by Förster
3 manuals	1928	Ivan Vïshnegradsky (Paris), built by Förster
2 manuals	1928	Hans Barth (New York), built by George L. Weitz of Baldwin
3 manuals (2 pianinos)	1931	Alois Hába, built by Förster

Young's *Well-Tuned Piano* (1964), which has been revived effectively since 1974 with a Bösendorfer piano. Serge Cordier has specialized in tuning pianos to equal temperament with justly tuned 5ths.

In the 20th century many temporary modifications have been made to the piano and new playing techniques applied to it. Isolated effects were required by Schoenberg in the Three Piano Pieces op.11 (1909), in which certain keys are silently depressed to raise the dampers and allow the strings to vibrate sympathetically, and Charles Ives, who in the 'Hawthorne' movement (1911) of his 'Concord' Sonata called for the use of a piece of wood 14¾" long for the playing of diatonic clusters; string glissandos, played by the fingers, are specified in Rued Langgaard's *Sfaerernes musik* (1918). More recently several composers have utilised small weights placed on or thin wedges inserted between keys in order to permit them to be silently depressed over longer durations.

The first composer systematically to explore the possibilities of modifying the piano was Henry Cowell. His innovations included the playing of chromatic and diatonic clusters (in *Adventures in Harmony*, ?1911, and *The Tides of Manaunaun*, ?1917) and glissandos across several strings or along single strings, executed with the fingers while the dampers are raised (*The Banshee*, 1925), plucking the strings (*Aeolian Harp*, *c*1923, and *Pièce pour piano avec cordes*, 1924), damping the strings with the fingers and small mutes, and playing them with hammers and plectra, to create what he termed the 'percussion piano' (*The Leprechaun*, *c*1925) and stopping the strings to alter the pitch or produce harmonics (*Sinister Resonance*, 1925). Sheets of paper are inserted between the strings in Satie's *Le piège de Méduse* (1914) and an upright piano modified in the same way is proposed by Ravel in his *Tzigane* (1924) and *L'enfant et les sortilèges* (1920–24) as an alternative to the original instrument intended by him, George Cloetens's 'luthéal', which adds two treble and two bass stops to a normal grand piano; these provide, separately and in combination, additional timbres resembling cimbalom, harpsichord and lute/harp, created by placing suspended metal bolts and additional felt dampers in contact with the strings. A similar system, in which thin brass tongues folded round strips of felt are placed between the hammers and the strings, was devised by Pleyel to make the sound of the piano resemble that of a harpsichord; it was used in a ballet (1926) by Gabriel Pierné and in Reynaldo Hahn's opera *Mozart* (1927). In the early 1930s in works for percussion ensemble William Russell specified simple preparations such as a cluster board, string glissandos, and strings plucked and struck by beaters.

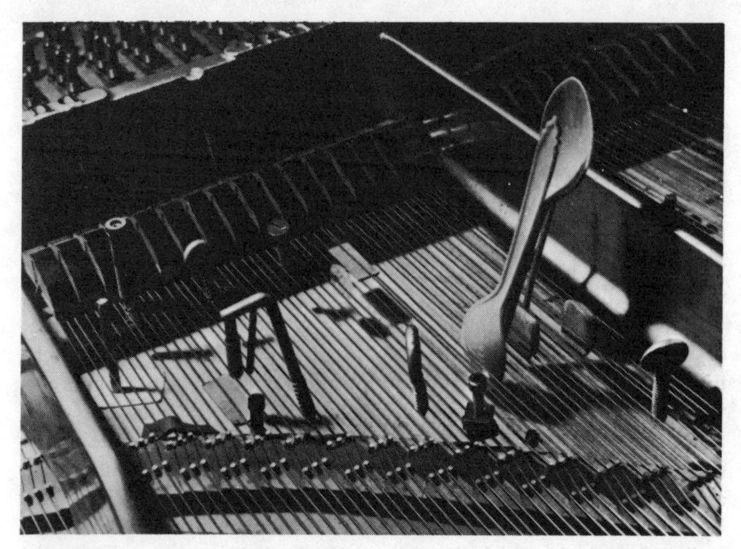

36. John Cage's prepared piano, 1940

The best-known of all piano modifications is John Cage's 'prepared' piano, devised in 1940 (not, as generally stated, 1938), in which a variety of objects is inserted between the strings, changing both timbre and pitch, to create a one-man percussion ensemble (see fig. 36); a range of different, more muted sounds is heard when the soft pedal is depressed. The prepared piano was the culmination of Cage's explorations of some of Cowell's ideas – the muting of strings both manually (*Imaginary Landscape no.1*, 1939) and with metal cylinders (*Second Construction*, 1940), and sweeping them with a stick (*First Construction in Metal*, 1939). Up to 1954 he wrote over 20, mostly solo, works for the prepared piano, some of them for dance performances. Simple preparations are also used in his works for 'string piano' of the early 1940s. (More recently a number of other composers and improvisers, including several from Japan, Hungary, Czechoslovakia, Poland and the Soviet Union, have used the prepared piano, mainly in ensembles; its inclusion for continuo-like colouration and punctuation in works by Arvo Pärt is especially notable.) Closely associated with Cage around 1940 was Lou Harrison, who devised the 'tack piano', in which thumb tacks or drawing pins are inserted into the hammers to create a metallic sound quality. This idea (known in German as the 'Reissnagelklavier' or the 'Reisszweckenklavier') was arrived at independently and applied mainly to upright pianos by other musicians, including the composers Henry Brant, Paul Dessau, Kagel, Wilhelm Killmayer and György Ráyki, as well as the honky-tonk pianist

Winifred Atwell; old and out-of-tune pianos, which produce a similar effect, have been called for (usually in theatrical contexts in connection with 1920s jazz or other popular musics), notably by Alban Berg, Max Brand, Peter Maxwell Davies, Karl Aage Rasmussen and Irwin Bazelon. The 'percussion piano' of Cowell was further developed by Lucia Dlugoszewski as the 'timbre piano' (1951), while Annea Lockwood and Hans-Karsten Raecke have developed their own, somewhat different approaches to preparing pianos.

The piano has been the subject of many modifications besides that of preparation, and considerable use has been made of different methods of playing the strings, frame and case, both with fingers and various implements, in works by Kagel, Cage, Orff, Lukas Foss, Ben Johnston and others. One feature of the 'timbre piano' is the use of small 'bows' directly on the strings for creating sustained sounds, and a similar approach has been adopted in several compositions since 1972 by Curtis Curtis-Smith, while since 1977 Stephen Scott has created a series of works in which the strings of a single instrument are bowed by an ensemble of about ten players with various flexible and rigid 'bows', as well as solo pieces in which silently depressed keys raise the dampers from the electromagnetically excited strings. Pianos with the action removed so that they must be played like a cimbalom have been specified by Peter Maxwell Davies and Denis ApIvor, and the strings are struck with T-shaped 'cluster-sticks' in works by David Bedford, Davide Mosconi and Sakis Papadimitriou, who specializes in playing entirely inside the piano. Foss, Bedford, George Crumb, Xenakis and others have called for undamped piano strings which vibrate sympathetically when other instruments are played nearby.

CHAPTER TWO

Piano Playing

The history of piano playing is tied to a great many factors: the development of the instrument, the evolution of musical styles, shifts in the relationship of the performer to the score, the rise of virtuosity, the idiosyncrasies of individual artists, changes in audience tastes and values, and even socio-economic developments. On a more practical level piano playing is concerned primarily with matters of touch, fingering, pedalling, phrasing and interpretation. Even a discussion limited primarily to these can point out only the major signposts along the two and a half centuries of the instrument's existence. Much of the lore surrounding the history of piano playing belongs more properly to the realm of anecdote or even myth than to scholarship; much work in this area remains to be done.

1. CLASSICAL PERIOD

The earliest performers brought with them well-established techniques for playing the harpsichord and clavichord, both of which were essentially domestic instruments in spite of their cultivation at leading courts throughout Europe. The best international keyboard repertory required considerable agility, dexterity and coordination, but minimal strength. With a maximum range of five octaves, coupled with longstanding resistance on the part of composers to the fully chromatic use of the keyboard (embraced only by J. S. Bach), there were inherent limits to the musical and technical demands a composer might make upon a player.

Much emphasis has been placed upon the similarity of the early fortepiano to both the clavichord and the harpsichord. There exist parallels in case and in soundboard construction; but as far as the performer was concerned the piano imposed a set of new demands. The various escapements introduced as early as Cristofori allowed the pianist to exert more downward pressure than was feasible on a clavichord. The fortepiano, however, was without the resistance encountered in pressing a plectrum past a string; its dip was correspondingly shallower. While the dynamic range of the new instrument was greater than that of a clavichord, it could not achieve the clavichord's various gradations of *piano*, and its maximum volume was still less than that of a well-quilled harpsichord. The special skills required for playing the piano are acknowledged obliquely in C. P. E. Bach's *Versuch* (1756): 'The more recent fortepiano, which is sturdy and well built, has many fine qualities, although its touch must be carefully worked out, a task that is not without its difficulties'. It is known that both Carl Philipp Emanuel and his father had access to the Silbermann pianos at the court of Frederick the Great in Potsdam, where the former was employed, but apart from Johann Sebastian's suggestions for improving the action on his visit in 1747 there is no documentation of his performances on the new instrument. Hence for the first six decades or so after its invention the piano co-existed with its more established rivals. Marpurg's *Anleitung* (1755) treats keyboard instruments as a family with broad performance skills in common. Even Türk's *Clavierschule* (1789) – cited by Beethoven in a conversation book as late as March 1819 – is directed as much at clavichordists as pianists. Until one instrument came to be preferred by composers and players alike, it was not economically feasible to aim a method book at a specialized audience. It is probably safe to assume that a still hand and an even touch remained the primary objectives of keyboard players until well after the death of J. S. Bach.

The persistence of these virtues is displayed in a letter Mozart wrote to his father from Augsburg in October 1777, wherein he criticized in biting fashion the playing of Stein's little daughter, Nannette, presumably on one of the maker's new fortepianos:

> When a passage is being played, the arm must be raised as high as possible, and according as the notes in the passage are stressed, the arm, not the fingers, must do this, and that too with great emphasis in a heavy and clumsy manner.... When she comes to a passage that ought to flow like oil and which necessitates a change of finger ... she just leaves out the notes, raises her hand, and starts off again quite comfortably.... She will not make progress by this method ... since she definitely does all she can to make her hands heavy.

Mozart's rival Clementi still admonished his pupils in his treatise of 1801 to hold 'the hand and arm ... in an horizontal position; neither depressing nor raising the wrist.... All unnecessary motion must be avoided'. Similarly, Dussek (1796) counselled the student 'never [to] displace the natural position of the hand'. Although Beethoven told Ries that he had never heard Mozart play, Czerny reported otherwise, attributing to Beethoven the observation that Mozart 'had a fine but choppy [*zerhacktes*] way of playing, no ligato'. This remark must be understood against the background of the gradual shift from non-legato to legato that had its beginnings in the high Classical period. Nevertheless, the keyboard music of Beethoven supplies the most imaginative examples of non-legato in the first quarter of the 19th century. In spite of his own legendary virtuosity and gift for improvisation, it is hard to form a coherent picture of Beethoven's performing style from contemporary reports. According to one of the best-known accounts, that by Carl Czerny, 'his bearing while playing was masterfully quiet, noble and beautiful, without the slightest grimace. . . . In teaching he laid great stress on a correct position of the fingers (after the school of Emanuel Bach)'. But Czerny appears to contradict himself in reporting further that Beethoven's 'playing, like his compositions, was far ahead of his time; the pianofortes of the period (until 1810), still extremely weak and imperfect, could not endure his gigantic style of performance'. And, according to Beethoven's biographer Schindler, 'Cherubini, disposed to be curt, characterized Beethoven's pianoforte playing in a single word: "rough"'.

Whether Beethoven performed it himself or not, it is certain that works like the 'Hammerklavier' Sonata op.106 demanded far greater technical resourcefulness (including participation of the full arm) than anything written before 1818. The last articulate spokesman for the conservative Viennese tradition was Hummel, whose *Anweisung* (1828) emphasized 'ease, quiet and security' of performance. In order to realize these goals, 'every sharp motion of the elbows and hands must be avoided'. Nevertheless, Hummel consolidated many of the innovations in fingering that had been adopted by Beethoven and others. More than 60 per cent of his method is devoted to this subject, with great stress on the pivotal importance of the thumb. Along with his own music Hummel advocated serious study of J. B. Cramer's *Studio per il pianoforte* (1804–10) and Clementi's *Gradus ad Parnassum* (1817–26), two of the first systematic surveys of keyboard technique. Although Cramer's goal of the absolute equality of the ten fingers was eventually abandoned, his studies were recommended enthusiastically by composers with aims as diverse as Beethoven, Schumann and Chopin. The heavier, more resonant (and less clear) English instruments preferred (and, in Clementi's case, manufactured) by English and French performers are compared without prejudice by Hummel with the lighter, transparent Viennese instruments. The gradual domination of the English type, including the eventual adoption of the double escapement patented by Erard in 1821, exercised a profound influence on the development of piano playing in the second half of the century.

2. ROMANTIC PERIOD

The dawn of Romanticism in the 1830s brought with it the specialization that produced a breed of pianists who were to dominate the salons and concert halls of Europe for the next 80 years. Although the number of amateur pianists continued to grow, the keyboard became increasingly the realm of the virtuoso who performed music written by and for other virtuosos. It is no accident that two composers on the threshold of the new movement, Weber and Schubert, each wrote a great deal of highly original piano music but were also highly original orchestrators, while two full-blooded Romantics of the next generation, Chopin and Schumann, have their achievements more clearly bounded by the capabilities and limitations of the piano. Weber was an accomplished pianist, but both he and Schubert dreamt of success in opera; Chopin became a highly polished virtuoso, while Schumann tried to become one. Among Romantic composers, some shunned or showed little interest in the piano (Berlioz, Verdi, Wagner), and others lived from its extraordinary powers, both as performers and teachers (Chopin, Liszt, Thalberg). This division helps to explain the intense interest after Beethoven's death in developing a range of sonorities for the solo piano that could be compared to an orchestra. Perhaps the most colourful example of this concern is the account by Charles Hallé of a concert he attended in Paris in 1836:

> At an orchestral concert given by him and conducted by Berlioz, the 'Marche au supplice', from the latter's *Symphonie fantastique*, that most gorgeously orchestrated piece, was performed, at the conclusion of which Liszt sat down and played his own arrangement, for the piano alone, of the same movement, with an effect even surpassing that of the full orchestra, and creating an indescribable furore.

The problems of studying piano playing are even more formidable over the Romantic era than over its beginnings. There are several reasons for this. In spite of the proliferation of method books by such artists as Moscheles, Herz and Kalkbrenner, none of the most innovatory contributors to 19th-century pianism (Schumann, Mendelssohn, Chopin, Tausig, Liszt, Brahms and Leschetizky) compiled similar guides. Chopin left behind the barest torso of a method book, apparently prompted largely by financial considerations and perfunctory in all but two respects. The closest testimonial in the case of Liszt is the largely neglected *Liszt-Pedagogium* (1901), assembled by Lina Ramann with fellow-pupils including August Göllerich.

Even more exasperating than the lack of guidance from the major performers themselves is the imprecision of the accounts in an age that worshipped flights of poetic fancy. From a novelist like George Sand one might expect the following description of Liszt at the keyboard:

> I adore the broken phrases he strikes from his piano so that they seem to stay suspended, one foot in the air, dancing in space like limping will-o'-the-wisps. The leaves on the lime trees take on themselves the duty of completing the melody in a hushed, mysterious whisper, as though they were murmuring nature's secrets to one another.

But the description of a professional musician like Hallé is scarcely of greater value:

> One of the transcendent merits of his playing was the crystal-like clearness which never failed him for a moment, even in the most complicated and, for anybody else, impossible passages.... The power he drew from the instrument was such as I have never heard since, but never harsh, never suggesting 'thumping'.

Nor is Schumann's comment to Clara: 'How extraordinarily he plays, boldly and wildly, and then again tenderly and ethereally!'. Mendelssohn is only slightly more helpful:

> [Liszt] plays the piano with more technique than all the others ... a degree of velocity and complete finger independence, and a thoroughly musical feeling which can scarcely be equalled. In a word, I have heard no performer whose musical perceptions so extend to the very tips of his fingers.

In the case of Chopin, the other revolutionary of Romantic piano playing, the ground is slightly firmer. It seems astonishing that, even as a fresh arrival in Paris, he could make the following remark:

> If Paganini is perfection itself, Kalkbrenner is his equal, but in a quite different sphere. It is difficult to describe to you his 'calm' – his enchanting touch, the incomparable evenness of his playing and that mastery which is obvious in every note.

Certain characteristics of Kalkbrenner's conservative style lingered, as in Chopin's advice to his young niece Ludwika to keep the 'elbow level with the white keys. Hand neither towards the right nor the left'. He staked out a more individual position in the tantalizing fragment of a piano method, owned and transcribed by Alfred Cortot (now in the Pierpont Morgan Library, New York; see J. J. Eigeldinger: *Chopin vu par ses élèves*, Neuchâtel, 1970, 2/1979):

> Provided that it is played in time, no one will notice inequality of sound in a rapid scale. Flying in the face of nature it has become customary to attempt to acquire equality of strength in the fingers. It is more desirable that the student acquire the ability to produce finely graded qualities of sound . . . The ability to play everything at a level tone is not our object. . . . There are as many different sounds as there are fingers. Everything hangs on knowing how to finger correctly. . . . It is important to make use of the shape of the fingers and no less so to employ the rest of the hand, wrist, forearm and arm. To attempt to play entirely from the wrist, as Kalkbrenner advocates, is incorrect.

Chopin recommended beginning with the scale of B major, 'one that places the long fingers comfortably over the black keys. . . . While [the scale of C major] is the easiest to read, it is the most difficult for the hands, since it contains no purchase points'. Although Hummel is cited by Chopin as the best source for advice on fingering, his own contributions to this area were bold and innovatory. The 27 studies composed in the decade between 1829 and 1839 (including three for Fétis and Moscheles's *Méthode des méthodes*) are a manifesto for techniques still in widespread use. While Cramer, Clementi and Hummel all include exercises based on arpeggios, Chopin extended their comfortable broken octaves to 10ths and even 11ths in his op.10 no.1; in spite of the easily imagined difficulties of high-speed execution he wrote to the strength of the hand, avoiding, for example, the weak link between the third and fourth fingers. The 'Black-Key' Etude

Ex.1 Chopin: Study in G♭, op.10 no.5

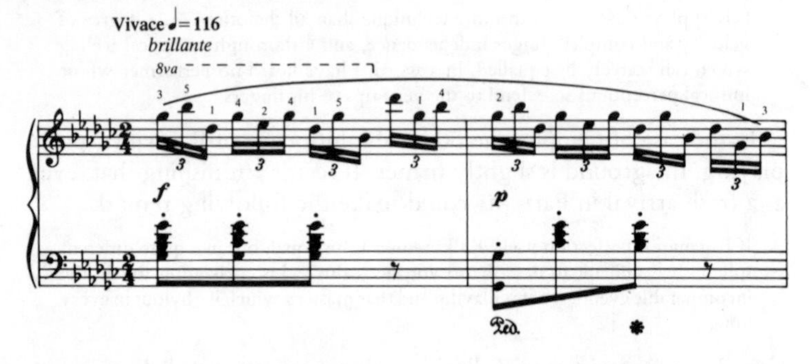

op.10 no.5 teaches the thumb to be equally at home on black or white keys (ex.1). The study in octaves, op.25 no.10, demands the participation (forbidden by Kalkbrenner) of the entire arm. Chopin provided fingering more frequently than almost any other 19th-century composer, adding them not only to autographs and copies but into editions used by students such as Jane Stirling.

Although Liszt's earliest efforts at technical studies were contemporary with those of Chopin, his own 'transcendental' studies, not published in their final form until after the latter's death, are repeatedly influenced by Chopin's example. The necessity for full involvement of the arm is readily evident from Liszt's fingerings in passages such as ex.2, from the sixth of the Paganini Studies. Brahms, who wrote two

Ex.2 Liszt: Paganini study no.6

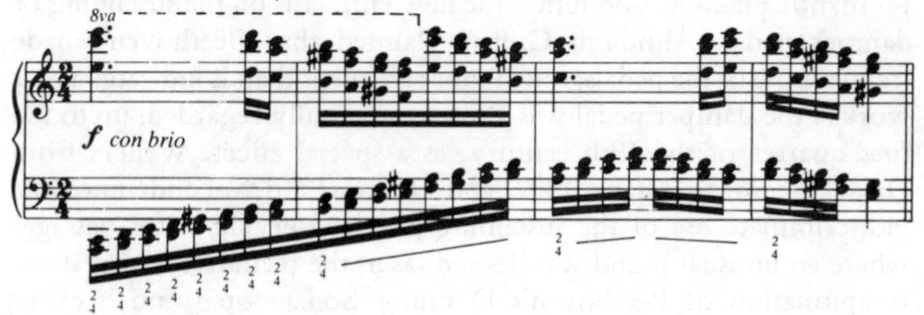

sets of variations on the theme of Paganini's A minor Caprice, favoured extensive cross-rhythms and metric shifts in his keyboard music. His specific contributions to piano technique are summarized in the *51 Übungen* (1893), which feature large leaps, sudden extensions and equally sudden contractions, and the passing of the fifth finger (i.e. the whole hand) over the thumb. This last device is employed freely in both hands of his last piano piece, the Rhapsody op.119 no.4 (ex.3).

Ex.3 Brahms: Rhapsody in E♭, op.119 no.4

Because of the lesser leverage available in the actions of the Viennese pianos that Brahms preferred, much of this technical expansion was accomplished on instruments with markedly greater resistance than that of present-day grands. Although data have been published purporting to show a steady increase in resistance through the 1870s, followed by a fall in the 20th century, much more extensive and reliable information will be needed before generalization about the relative touch of differing instruments will be possible. In the second half of the 19th century the only constant in this area was probably variety.

The single most important development in the sound of the Romantic piano was doubtless the new emphasis on the sustaining or damper pedal. Although Czerny claimed that Beethoven 'made frequent use of the pedals, much more frequent than is indicated in his works', the damper pedal was almost universally regarded, up to the first quarter of the 19th century, as a special effect. Writers from Dussek (1796) to Adam (1802) and Hummel (1828) condemned the indiscriminate use of the sustaining pedal, reserving it for passages where an unusual sound was desired (as in the recitative added at the recapitulation of Beethoven's D minor Sonata op.31 no.2; ex.4). Directions for raising the dampers were transmitted in very individual ways by Romantic composers; Schumann was among the first to specify simply 'Pedal' at the head of a passage or movement, while Chopin generally supplied precise and detailed instructions (frequently ignored or suppressed by his 19th-century editors). It is seldom clear whether Chopin intended those passages not marked (such as all but the first three bars of the opening section of the F major Ballade op.38) to be played without the damper pedal, or whether it was to be added as general colouring at the performer's discretion.

Liszt's teacher Czerny was one of the first to exchange public performing for full-time instruction, but a dominant specialist teacher did not emerge until after mid-century in the person of Theodor Leschetizky, who numbered among his pupils Paderewski, Gabrilovich, Schnabel, Friedman, Brailowsky, Horszowski, Moiseiwitsch and many more who achieved international fame. Although it became fashionable to speak of the 'Leschetizky method', Leschetizky himself steadfastly refused to freeze his views into print. In searching for the kernel his student Moiseiwitsch observed that 'above all there was his tone. No-one had a tone like his. He never taught us any "secret" there; one just picked up something of the lustre from him'. Perhaps an even greater contribution was Leschetizky's detailed and painstaking approach to the study of repertory, a tradition still pursued in countless master classes. Although his English successor Tobias Matthay (of German parentage) produced many books on piano playing, their tortuous language required explications by students (e.g. A. Coviello: *What Matthay meant*, 1948). Matthay's em-

Ex.4 Beethoven: Sonata in D minor, op.31 no.2, first movt

phasis on muscular relaxation and forearm rotation was valuable as far as it went but has needed modification in the face of more detailed physiological investigations like those of Otto Ortmann (1929). Ortmann's research led him to the not surprising conclusion that the most efficient playing requires a judicious balance between muscular relaxation and tension.

Few editors of piano music before 1930 approached their task with the reverence for the composer's intentions found in Schenker's 'Erläuterungsausgaben' (1913–21) of the late Beethoven sonatas. It was not only customary but expected that an editor would add his interpretative suggestions to those provided by the composer, rarely bothering to distinguish between the two. Since most 19th-century editors were themselves active performers who frequently claimed direct association with the composer of the repertory being edited, an interventionist attitude was inevitable. The most frequent text changes were the addition of articulation slurs in the music of Bach and Handel – then considered a regular part of the piano repertory – or the exchanging of articulation slurs (especially in the Viennese repertory from Haydn to Schubert) for longer phrase markings. The wholesale addition of dynamic and pedal indications was equally acceptable. In performance the pianist reserved the right to introduce further changes, perhaps restricted to a few discreet octave doublings but perhaps also extending to the interpolation of embellishments and cadenzas. Although it is known that both Beethoven and Chopin objected to such practices, the practices flourished. The most gifted practitioner may have been Liszt, who did not regard even Chopin's music (as the latter bitterly noted) as sacrosanct. Nevertheless, Chopin himself occasionally interpolated embellishments and cadenzas into his music, as shown in an annotated version of op.9 no.2, which shows a variant of the cadenza and an added flourish to the final bar. In later years Liszt renounced his earlier habits, crusading relentlessly over the tinkling of salon music for the acceptance of works by Beethoven, Schubert, Berlioz and others.

The recent vogue for 'Urtext' editions has reaffirmed the importance of the composer in the chain leading to actual performance, but an enthusiasm for textual purity can prove dangerous when accompanied by naïvety about the performing conventions and traditions known to contemporary players. In general, variety in articulation persisted much longer than is usually acknowledged, proving essential not only in the music of Haydn and Mozart but also in that of Schubert and Chopin. Romantic composers handled the issues of phrasing and articulation in highly individual ways, frequently alternating between the two types of notation within the same movement, section or even phrase. Because of the complex relationships among primary sources it is rarely a simple matter to establish an 'Urtext', as the comparison of two such editions of almost any work will prove. The realization that not only Mozart and Beethoven but also Chopin and Liszt played on instruments quite different from our own raises the nagging question of whether a modern performer on a modern instrument should attempt to adapt his playing style to that of the earlier piano or should feel free to make changes he feels are necessitated by intervening developments. Indeed, until a significant number of 19th-century instruments by such makers as Graf, Streicher, Broadwood, Bösendorfer, Pleyel, Erard and Steinway are restored to concert condition, there can be little more than speculation as to how they actually sounded, or even whether it would be desirable to include them as a regular part of concert life. Who would advocate playing keyboard music before Dussek (supposed to have been the first to turn his right profile to the audience) with his back to his listeners? Should music before Liszt (the first to perform regularly in public from memory) be played with the music and a page-turner? The renewed interest in historical performance will not make the performer's task less complex; it both increases his options and his obligation to become fully informed.

3. 20TH CENTURY

The development of piano playing in the 20th century received its major impetus from Claude Debussy, who took up where Chopin had left off five decades earlier. Unlike most 19th-century piano composers, Debussy was no virtuoso (few accounts of his playing, and only a fragmentary recording accompanying Mary Garden in a scene from *Pelléas*, survive), but he was on intimate terms with the instrument to which he returned again and again. His piano music is an eclectic blend of Couperin and Chopin (the keyboard composers he admired most) combined with daring new harmonies and textures. The *Suite pour le piano* (1901) proved a landmark in 20th-century pianism, skilfully blending three centuries of keyboard tradition. It should be noted that Debussy achieved his finely graded pedal effects (never specified but always an integral part of the texture) without the benefit of the middle, 'sostenuto' pedal found on most modern concert instruments. The capstone to Debussy's piano writing is the set of

Ex.5 Debussy: Etude X (*Pour les sonorités opposées*)

twelve *Etudes* (1915), fittingly dedicated to Chopin. Beginning with the spoof on 'five-finger exercises' through a chord study, these essays prepare the performer not only for the rest of Debussy's piano music but for much of the keyboard music that followed. Unlike the Romantic composers who cultivated a homogeneous blend, Debussy revelled in 'opposed sonorities', as in his *Etude* of that name (ex.5). In spite of notational fastidiousness in matters of dynamics and phrasing, he elected in the preface of the *Etudes* to grant the performer complete freedom in another important area:

> To impose a fingering cannot logically meet the different conformations of our hands. . . . Our old Masters . . . never indicated fingerings, relying, probably, on the ingenuity of their contemporaries. To doubt that of the modern virtuosos would be ill-mannered. To conclude: the absence of fingerings is an excellent exercise, suppresses the spirit of contradiction which induces us to choose to ignore the fingerings of the composer, and proves those eternal words: 'One is never better served than by oneself'. Let us seek our fingerings!

The cross influences between Debussy and Ravel may never be entirely sorted out, but it is at least clear that Ravel remained more drawn to the cascades of virtuosity inherited from Liszt. His special fondness for rapid repeated notes (as in *Gaspard de la nuit*) presupposes a crystalline control of touch and nuance essential to all of his music. Although also influenced by Debussy, Bartók travelled an increasingly individual path, beginning with the *Allegro barbaro* of 1911. He is noted for the spiky dissonance that punctuates his keyboard music, but it is too often forgotten that his own playing – both from the recollections of contemporaries and the evidence of numerous sound recordings – was infused with great elegance and rhythmic subtlety. Nevertheless, his frank exploitation of the percussive capabilities of the piano helped pave the way for the experiments with 'prepared' pianos first introduced in Cage's *Bacchanale* (1940) and embraced by many composers since. The placing of small wedges of india-rubber or other materials between the strings to modify the sound is curiously analogous to the mechanical means used in the harpsichord of two centuries earlier. Other means of tone production, such as tapping the case or the soundboard, have also been added. No standardized notation for transmitting these directions has evolved, varying not only from composer to composer but from work to work by the same composer. These idiosyncratic developments, along with the new interest in historical performance, have helped mitigate the increasing postwar homogenization in the interpretation of the standard repertory.

4. JAZZ PIANO PLAYING

As an improvised art, which is often highly complex, jazz places special demands on piano technique, and jazz pianists have evolved a brand of virtuosity quite distinct from that of the classical tradition. Jazz and blues pianists do not generally set out to acquire an all-embracing technique capable of handling a wide-ranging body of literature; they concentrate instead on mastering a few technical problems which pertain to their particular style, personality and individual interests. Within these deliberately narrow confines their technical attainments have been quite remarkable, such as the perfect rhythmic separation of the hands required by the boogie-woogie style, the rapid negotiation of wide left-hand leaps in the stride style, or such individual traits as Teddy Wilson's gentle emphasis of inner counterpoints with the left thumb; even classical pianists have difficulty handling these technical problems without sacrificing jazz propulsion or 'swing'. Thus pianists of quite limited technique, such as

Jimmy Yancey, Thelonious Monk and Horace Silver, have developed distinctive and inventive jazz styles whereas piano virtuosos, such as Friedrich Gulda, André Previn or Peter Nero, have not been as successful.

Jazz piano playing evolved early in the 20th century from several separate strands, the most important being ragtime, which was easily within the grasp of the amateur pianist. Its characteristic features – a march-like accompaniment pattern in the left hand against syncopated broken chords in the right – became more technically complex in the 1920s with the Harlem stride school. In a spirit of keen competition its members deliberately set out to dazzle listeners, and especially colleagues, with the speed and daring of their technique. One feature that became almost a fetish was the 'solid left hand', where three-octave leaps at rapid tempo were not uncommon and octaves were regularly replaced by 10ths. By contrast, the right hand played light and feathery passage-work with rapid irregular 3rds and pentatonic runs (fingered 3–2–1–2–1). The finest jazz technician, Art Tatum, was especially adept at integrating the hands in rapid passage-work and commanded the admiration of Horowitz; few jazz pianists have been able to match his virtuosity, the only exception perhaps being Oscar Peterson.

A contrasting style arose in the late 1920s with the work of Earl Hines. His 'trumpet style' translated many of the inflections of jazz trumpeting to the right hand of the piano in the form of irregular tremolandos, clusters and punched chords and a thin texture with abrupt *sforzati* and cross-accents. Another development was the boogie-woogie blues style of the 1930s. Here an unwavering rhythmic pattern in the left hand was offset by irregular cross-rhythms in the right, necessitating an absolutely secure rhythmic separation of the hands. Though crude and homespun by the standards of Tatum and Hines, boogie-woogie nevertheless left its mark on later rhythm-and-blues and rock pianists.

In the 1940s, the 'bebop' style represented a radical rethinking and simplification of previous jazz piano playing. The rhythmic function of the left hand was taken over by the drums and bass of an ensemble and the pianist was left to spin out long lines of 'single-note' melodies (i.e. with one note played at a time) while outlining the harmonic progressions and 'kicking' the beat with sparse chords in the left hand. The emphasis was on a precise and mobile right-hand technique capable of sudden cross-accents, which were generally accomplished by a quick wrist staccato. The inevitable outcome of this approach was an extremely restrained sonority (the pedals were virtually ignored), yet the best bop pianists such as Bud Powell, Thelonious Monk and Horace Silver cultivated a readily recognizable and inimitable touch.

Key figures of the late 1950s to rediscover the different timbres of the instrument were Bill Evans and Cecil Taylor. Evans cultivated an understated technique consisting of blurred pedal effects, gentle tremolandos, careful spacing of notes in a chord ('voicing'), a fondness for low dynamic levels and implied rather than explicitly stated rhythms. Taylor, a pianist with a conservatory training, chose avant-garde art music as his starting point and pursued a physically demanding style with clusters, glissandos and palm- and elbow-effects such as those found in Stockhausen's later piano pieces. Like Evans, Taylor made use of the full tonal range of the instrument, but to completely different and more extrovert ends.

Jazz pianists today are usually trained in a sound classical technique and have a historical grasp of earlier jazz piano playing. This has led to interesting hybrids of classical and jazz technique, such as the work of Chick Corea and Keith Jarrett. The technical expertise of the players is considerable and almost encyclopedic in scope. The advent of the electric piano has brought a new array of technical problems, such as the handling of the bend bar and the manipulation of volume, wah-wah and other pedals; these have been particularly well mastered by Herbie Hancock and Josef Zawinul.

CHAPTER THREE
Pianists

1. LISZT AND HIS CONTEMPORARIES

Piano playing as we know it today starts with Franz Liszt (1811–86). There were, of course, great pianists before him, from Mozart to Beethoven, Clementi, Hummel, Cramer, Dussek, Moscheles and others, and they were the ones who established the lines of descent of the 19th century. Clementi taught Moscheles and Moscheles taught Mendelssohn; Hummel taught Thalberg; Beethoven taught Czerny who taught Liszt and Leschetizky. From the Liszt and Leschetizky studios came many of the supreme pianists of the century. Liszt taught, among many others, Tausig, Bülow, Lamond, Rosenthal, Ziloti, d'Albert, Menter, Sauer and Joseffy. Leschetizky taught Paderewski, Friedman, Gabrilovich, Moiseiwitsch, Hambourg and Schnabel. They in turn had their pupils. Separate lines of descent were established in France and Russia.

Liszt was supreme among pianists, the equivalent of Paganini on the violin, the first of the supreme piano virtuosos, the showman, the superstar. He had all the gifts: good looks, health, charisma, a remarkable musical mind, an infallible memory, and the ability to sightread anything, no matter how difficult, as though he had played it all his life. He had the entire known repertory at his fingertips. In 1839 he even invented the solo piano recital; before him, no instrumentalist gave a concert without assisting artists. During Liszt's lifetime the piano much as we know it today evolved. It was then that the Industrial Revolution brought with it the railway system, which now could transport touring musicians like Liszt all over Europe with relative ease. Franz Liszt was the right man in the right place at the right time.

His actual career as a concert pianist was short, from 1839 to 1847, but he never stopped playing the piano in public or teaching. He was by far the greatest pianistic force of the century. In his concerts he played not only his celebrated virtuoso pieces and arrangements but was also the first to play in public, with some degree of consistency, many Beethoven sonatas (including the last five), Bach's Goldberg Variations and selections from the '48', music by Schubert, Schumann and Mendelssohn, and a great deal of Chopin. He presented new ideas of technique, a transcendent virtuosity that none could match, and a kind of luminous sound that was considered a miracle by all who heard it.

Not until the last quarter of the 19th century did pianists appear who could stand comparison with him. Of Liszt's major rivals in his lifetime, Chopin, who influenced him, gave few concerts and never toured after settling in Paris in 1831. While Chopin developed new types of piano style, Liszt was the one who created modern virtuosity. His other great rival, Sigismond Thalberg, excited the public and was an elegant virtuoso stylist, with, however, a stereotyped method: he placed the melody in the centre of the keyboard (using the thumbs of both hands and the sustaining pedal to prolong the sound), ornamenting it with florid counterpoint, arpeggios and chords above and below. Neither he nor his music had much staying power.

If Liszt was the archetype of the artist-as-hero, the crowd-pleasing virtuoso, there was another school in which sobriety, idealism, adherence to the printed note and to the composers' intentions were governing factors. The leading representative of that school was Clara Wieck (1819–96), the wife of Robert Schumann. As she matured she dropped most of the salon, exhibitionistic music and other dross from her repertory and played only great music from Bach to Brahms. She also was the first prominent pianist to concentrate on chamber music, and appeared throughout the century with Joseph Joachim, the greatest classical violinist of his time. With him she played not only sonata programmes but, with his quartet, the masterpieces of the chamber literature. Other eminent contemporaries of Liszt's include Henri Herz (1803–88), whose elegant style was modelled on that of Moscheles, and Adolf Henselt (1814–89), a pupil of Hummel's who was noted as a Chopin interpreter and whose technique was characterized by the sustaining (without pedal) of widely arpeggiated figures and chords.

Starting with Hans von Bülow in the 1850s, the Liszt pupils began to make their impact. Bülow was followed by Carl Tausig (1841–71), and then by the extraordinary group of Liszt's last years headed by Moriz Rosenthal (1862–1946) and Eugen d'Albert (1864–1932). The peppery, didactic Bülow was an intellectual whose playing was accused of dryness. His severe programmes included the last five Beethoven sonatas and the Diabelli Variations (Clara Schumann avoided such monster works as this or the *Hammerklavier*) and he

37. Franz Liszt at a Graf piano with (from left to right) Alexandre Dumas, Victor Hugo, George Sand, Paganini, Rossini and the Countess Marie d'Agoult: painting (1840) by Joseph Danhauser

insisted on playing them to audiences all over the world, including the USA. Tausig, probably the greatest of Liszt's pupils, and a friend of Brahms as well as of Wagner, was considered the most flawless pianist of his time, and indeed may well have been the most flawless ever.

Of his late pupils, Liszt had the highest regard for d'Albert, whom he considered his most gifted pupil besides Tausig. D'Albert impressed with his fiery temperament, his seemingly unlimited technique and his breadth of musical vision. Bruno Walter regarded d'Albert as the greatest Beethoven player at the turn of the century, and wrote that his performance of the 'Emperor' Concerto was unparalleled. After a brief concert career, d'Albert turned to composition and let his skills evaporate; his recordings, none of them very good, obviously do not represent the 'little giant' who startled the world. Rosenthal, who did not record until he was in his 60s and thus past his prime, was considered the most brilliant technician of his day, superior even to Liszt 'in certain aspects of technique', as one of Liszt's pupils guardedly said. He excited the public with his thunderings, but (as his recordings testify) he also was capable of a *jeu perlé* almost approaching Josef Hofmann's. His playing had culture, style and musicianship behind it, and he must have been one of the most impressive pianists of the pre-World War I period.

2. THE LESCHETIZKY SCHOOL

A generation after the Liszt pupils, the pupils of Theodor Leschetizky (1830–1915) started taking over the piano establishment. So many important pianists emerged from this teacher that there was much talk about the 'Leschetizky system'. Leschetizky himself denied that there was such a thing. But it was a remarkable teacher indeed who could produce artists so great yet disparate as Paderewski and Artur Schnabel, Mark Hambourg and Benno Moiseiwitsch.

The first of the great Leschetizky pupils was Ignacy Jan Paderewski (1860–1941), the most fabulously successful pianist of all time. He was not, however, the most gifted pianist of his period. A child prodigy but poorly trained, he did not begin serious studies until he went to Leschetizky, who promptly put him on beginners' exercises. He showed remarkable perseverence and will-power: lacking the technical fluency of the wizards of the time, he produced his effects by other means, of which charisma was not the least. His rich, sensuous tone and nobility of conception seemed to make audiences forget his digital problems. His interpretations were the most Romantic of all the Romantics, full of left-hand-before-right effects and an unusually abundant rubato.

A Leschetizky pupil who attracted considerable attention in his time and who today, thanks to his recordings, is considered one of the supreme Romantic pianists, was Ignacy Friedman (1882–1948). Polish-born, he brought rhythmic qualities to the Chopin mazurkas that no pianist has duplicated. In addition Friedman had a stupendous technique and one of the richest sonorities of any pianist in history, as is evident in his recording of Chopin's E flat Nocturne (op.55 no.2): a performance that is a lesson in how to project a melody, how to keep it in motion through delicate tempo adjustments, and how to handle a bass line so that it all but counterpoints the melody.

Three major national schools in piano playing began to manifest themselves: a Slavonic school, a German school and a French school, each with its own characteristics. These continued until relatively recent times.

3. THE SLAVONIC SCHOOL

Anton Rubinstein (1829–94), active in St Petersburg, was the creator of the Slavonic school; later his brother Nikolay, in Moscow, supplemented his work. Before Rubinstein there was no such thing as a Russian pianist. But Anton Rubinstein turned out to be the pianist who captured the public imagination more than any performer after Liszt. He had a striking appearance – a shambling bear of a man with a leonine face. Although he possessed a colossal technique and a colossal sonority, he could be a sloppy pianist with numerous wrong notes. When he became excited at the keyboard anything could happen. There was, however, something elemental in his outsize playing: characterized by tremendous surges of energy, which could degenerate into pounding, it was never less than virile and grand and was also capable of ravishing *pianissimo* and colour effects. Rubinstein's programmes could run well over three hours. He would play Chopin's B flat minor Sonata as an encore, and follow it with the seven pieces of Mendelssohn's *Characterstücke*. When he toured the USA in the 1872–3 season he gave 215 concerts in 239 days. He many times gave a series of historical recitals, surveying the literature from Elizabethan composers up to Liszt in seven enormous programmes. He founded the St Petersburg Conservatory in 1861;

38. Josef Hofmann and Anton Rubinstein: painting by Charles E. Chambers

Nikolay founded the Moscow Conservatory five years later.

The Slavonic, or Russian, school can be described as warm-hearted, generous, extroverted, colourful and yet (judging from recordings of its exponents born in the 1870s) emotionally and rhythmically controlled. Surprisingly little rubato is used; even more surprisingly, again on the basis of recordings of pianists born in the 19th century, the German school used much more rubato than the Slavonic. Instead, the Slavonic school used constant fluctuations of tempo, with ritardandos to announce contrasting sections and accelerandos to emphasize dramatic points. In common with all Romantic pianism, the Slavs emphasized tonal beauty, and they never produced an ugly or percussive sound no matter what dynamics were employed.

The most important successors to Anton Rubinstein were such pianists as Sergey Rakhmaninov (1873–1943) and Josef Lhévinne (1874–1944), from Russia, and Leopold Godowsky (1870–1938) and Josef Hofmann (1876–1957), from Poland. Rakhmaninov and Hofmann, both active until close to their deaths, were among the greatest and most popular of the late Romantic pianists. They had certain traits in common with all the Romantics – gorgeous tone, a singing line, infallible rhythm, a manly sense of poetry that completely avoided sentimentalism, an ability to organize and move bass lines so that extra colour and even polyphony were added to the musical texture, and the ability to bring out inner voices (nearly always notated or implied by the composer and largely ignored today). Both had very much the same repertory – little Bach, Haydn or Mozart; a handful of Beethoven sonatas; and then the entire 19th-century literature with the exception of Brahms (though Hofmann played the Handel Variations).

There were, however, significant differences between their styles. Rakhmaninov was the exemplar of planning, and seldom varied his interpretations from year to year. Hofmann seldom played anything twice the same way; an improvisatory feeling was always present. Until the later years of his life Hofmann was one of the cooler Romantics, with a neo-classical style; some experts, used to the freedom of the older Romantics, at first considered him too tight. He was one of the first Romantics to insist on the supremacy of the printed note, though that did not prevent him from making subtle textual changes when he thought them necessary. He had one of the most remarkable mechanisms in pianistic history: his scales and trills were clearer and faster than anybody else's, his chords and octaves were crisp and perfectly placed, and the sheer finish of his pianism struck fear and envy into his colleagues.

Lhévinne too was one of the giants, an aristocratic artist with an enormous technique and command of colour as well as the most supple manner of delivery. Godowsky was considered the pianist's pianist. In public, it is said, he held back a little too much and his playing was considered somewhat cool. In his studio, however, pianists came to marvel at his flawless delivery and his phenomenal ability to juggle the contrapuntal strands of his own paraphrases on Johann Strauss and the Chopin études.

4. THE GERMAN SCHOOL

Idealistic, in the Hegelian sense, and much more severe than the Slavonic, the German school as taught in Leipzig, Berlin and Vienna always has been one with a high seriousness, avoiding superficial prettiness in favour of stringent musicianship. It concentrates largely on the Austro-German musical heritage (it is hard to think of any German pianist who has been recognized as a great Chopin player). Bülow and Clara Schumann were the first examples of the species. Ferruccio Busoni (1866–1924), Italian-born but half German, taught in Berlin for much of his life and can be considered an exponent of the German school. His specialities were Bach, Beethoven, Liszt and, at the end of his life, the Mozart piano concertos. Busoni must have been one of the most original of pianists, at his best in the heroic works of the repertory, such as the *Hammerklavier* or the big Liszt operatic paraphrases. His greatest pupil, Egon Petri, carried on the Busoni line.

Then came Artur Schnabel (1882–1951), who in his youth played everything but eventually discarded most of his repertory in favour of Bach, Mozart, Beethoven and Schubert. In some respects Schnabel remained a Romantic, not surprising considering his period and his teacher (Leschetizky); his interpretations had a controlled freedom in rhythm characteristic of the Romantics. Schnabel, however, was much stricter than most pianists of his day. He also was one of the first to go back to the sources, collating manuscripts and first editions of the music he played. A scholar as well as a great pianist, he purged Beethoven playing of its Romantic excrescences.

Contemporary with Schnabel was Wilhelm Backhaus (1884–1969), another great exponent of the German style. A much stronger technician than the sometimes erratic Schnabel, who never worried much about wrong notes, Backhaus offered monolithic performances of Beethoven and Brahms. He did not look for the colouristic effects inherent in the Slavonic style; rather he concentrated on structural elements in the music. It was Olympian but rather aloof. An anomaly was Walter Gieseking (1895–1956), who was considered the leading Debussy and Ravel player of his time; otherwise his repertory was largely Austro-German. His memory was fabulous. He first played the Beethoven sonata cycle when he was 16 years old, by which time he had already memorized most of Bach, all of Beethoven, Chopin and Schumann, much Mendelssohn and Schubert and, as he wrote in an autobiographical sketch, very little of Liszt and 'absolutely nothing' of Brahms. The Brahms deficiency he was to rectify. In more recent years, such pianists as Edwin Fischer (1886–1960), Rudolf Serkin (*b*1903), Wilhelm Kempff (*b*1895) and Claudio Arrau (*b*1903) have been outstanding examples of the German tradition of piano playing. Arrau, Chilean-born but German-trained, has the most catholic repertory of any of his colleagues: he has given Bach, Beethoven, Mozart,

39. Alfred Cortot

Schubert and Schumann cycles, plays a considerable amount of Chopin and Brahms, all the piano music of Debussy and Ravel and later 20th-century pieces.

5. THE FRENCH SCHOOL

For a good part of the century French pianism, centred at the Paris Conservatoire, was conditioned by the teachings of Pierre-Joseph-Guillaume Zimmermann and Antoine François Marmontel, who between them headed the piano department there from 1821 to 1887. Both were pianists of the old school, with neo-classical leanings and the kind of fluent but shallow technique represented by such salon pianists as Henri Herz (who also taught at the Conservatoire). Their tradition was carried on by Isidore Philipp. Their repertory concentrated on French music and also Chopin and Liszt (two composers considered virtually French nationals because of their long residence there). Until recent years, few French pianists regularly played Beethoven and almost none of them Brahms, and to this day French pianists play elegantly 'on top of the keys', with a rather percussive sound and a restricted sonority, faster tempos than pianists elsewhere, and with a civilized urbanity rather than deep probings into musical meaning. The French style was carried to its height by Raoul Pugno (1852–1914). His recordings (he made a large number around 1905) demonstrate an incredible fleetness of finger, a limpid style and a logical approach that is always to the musical point. Edouard Risler (1873–1929) was the first French pianist to specialize in Beethoven.

But it was Alfred Cortot (1877–1962), the greatest of all French pianists, who broke the mould. Cortot stood a little outside the French tradition. As a young man he had become a Wagnerian, worshipped at Bayreuth, and returned to Paris in 1902 to conduct the first performance there of *Götterdämmerung*. He had heard German pianists and he took many elements of their style into his playing. His sonority was richer than that of most other French pianists, his interpretations more introspective and idiosyncratic, his keyboard approach *sui generis*. He used a good deal of rubato in a very personal, effective manner. He had a technique that could handle anything in the repertory, as witness his majestic performance on records of the Liszt Rhapsody no.11. But such was his teaching schedule, his editorial work and other activities, that he had little time to practise. Any Cortot performance has an unusual share of missed notes, though they hardly matter in view of the nobility, personality and charm of his playing. His repertory was Romantic. He played little Bach, no Mozart and relatively little Beethoven. After that he seemed to have everything in his fingers, including the then new music of Debussy and Ravel. He was especially respected for his Chopin and Schumann playing. Until Artur Rubinstein came along he made more records than any other pianist; he was in the studios from about 1910 until almost the day of his death.

6. AFTER WORLD WAR I

After the end of World War I, a general retreat from Romanticism began. The new composers – Stravinsky, Prokofiev, Hindemith, Bartók – looked on the piano as a percussion instrument and wrote accordingly. It was in the 1930s and 1940s that their strictures were to take hold. But in the 1920s their music was generally ignored by the pianists born in the 19th century or the early 1900s. Those pianists were still trained in the Romantic conventions.

The years immediately following the war saw the emergence of the new British school, headed by Myra Hess (1890–1965), Clifford Curzon (1907–1982) and Solomon [Cutner] (*b*1902). It was an eclectic

40. Vladimir Horowitz, 1986

school in which all three of those fine pianists seemed to combine all European schools into a civilized, cultured amalgam – urbane, refined, elegant, seldom passionate. Perhaps this was reflective of the national temperament. Their playing combined taste and knowledge. For the most part their repertory centred on the Austro-German classics; Hess and Solomon also played much Chopin.

Of the three, Solomon, whose career was cut short at the height of his powers by a stroke, was the most remarkable. He had a perfect technique, a ravishing sound and a superior musical mind. His was playing suffused with intelligence and love, backed by exquisite taste and beautifully proportioned conceptions. The Russian-born Benno Moiseiwitsch (1890–1963) may also be considered British in view of his long residence in London. Moiseiwitsch is one of the more underestimated pianists. In the 1920s he seemed the natural heir of Hofmann and Rakhmaninov. He had a colossal technique that made everything sound easy, and used that technique to make honest music. Even in virtuoso showpieces, such as the Weber-Tausig *Invitation* or the Strauss-Godowsky *Fledermaus*, he never lost tonal or musical control. His interpretations were natural and unforced. Many consider his recording of Brahms's Handel Variations the most elegant and poetic ever made.

In 1925 the young Vladimir Horowitz (*b*1904) came out of Russia to startle the West. Here was Anton Rubinstein reincarnate, though a Rubinstein with perfect marksmanship. Horowitz remained an infallible technician, unlike Rubinstein, and would go through the most difficult music without ever missing a note. There was more than technique to his playing, however. Horowitz produced a sonority unique in the annals of pianism. It was not only colourful; it also had immense volume without being noisy. At Horowitz's first performance of the Tchaikovsky B flat minor Concerto in Hamburg, the stunned conductor, Eugen Pabst, left the podium after the opening chords to watch the pianist's hands and marvel at the immense volume that was being produced by the slim, quiet figure at the keyboard.

From the beginning, Horowitz had an electric quality to which audiences responded as to no other pianist. And for years he was critically untouchable. But a new generation of critics, trained in a style that did not respond to Horowitz's kind of Romanticism, started to attack him. They called his playing unmusical; they derided his *affetuoso* style in which, they claimed, the musical line was teased and even tortured out of recognition. There also was in his playing a neurotic quality that disturbed some listeners. Professional pianists, on the other hand, seemed universally to admire and indeed revere

Horowitz. They responded to his immense craft, to the colour he was able to achieve, to the combination of power and charm in his playing. Nobody ever claimed that he was an intellectual giant, but to many he was as much the legendary pianist as to violinists Jascha Heifetz was the legendary violinist (no matter what the critics thought).

The only pianist after the deaths of Hofmann and Rakhmaninov who rivalled Horowitz in public affection was Artur Rubinstein. Although Rubinstein was considerably the senior of Horowitz, being born in 1886, it was not until the 1930s that his career exploded, thanks in part to some superb recordings (the Tchaikovsky B flat minor Concerto, the Chopin scherzos and other Chopin works). If the Horowitz style disturbed many knowledgeable listeners, Rubinstein's was universally admired. His playing had an indescribable *joie de vivre*. It was happy, lucid, unneurotic, tasteful, intelligent and urbane, backed by a golden sound and strong fingers. Rubinstein never was an impeccable technician; he was too busy enjoying life to pay much attention to practising. And yet he was capable of stirring virtuoso feats when he put his mind to it. He had one of the longest careers of any pianist, playing in public from childhood until about five years before his death in 1981 at the age of 95.

7. AFTER WORLD WAR II

In the decades following the end of World War II an international movement made itself felt. National schools of composition and performance tended to become smoothed-out and interrelated. Just as it was difficult to tell the difference between the serial compositions of French, German and American composers, so it was difficult to tell the difference between American, German and, more recently, Russian pianists.

It was not yet that way in 1956, when the first contingent of Russians appeared in the West after an absence of many years. Emil Gilels (1916–85) and Sviatoslav Richter (*b*1915), not to mention the violinists, cellists and singers who followed them, impressed as brilliant musicians whose work was rather provincial. That is, it had been untouched by the varying aesthetic, musicological and intellectual influences that beat down as a matter of course on any Western artist. That lasted only a few years. As more Russian musicians played in the West they exchanged ideas with their colleagues everywhere, and their playing became much more cosmopolitan.

Gilels was a healthy, extroverted artist with a virtuoso technique which he never used for the purposes of mere display. He had a serious approach to music and was one of the steadiest, most dependable pianists on the circuit. Richter's career was unusual. As a youth he spent more time reading opera scores than practising repertory, and it was as a conductor that he started his musical life. Not until he was about 20 did he enter the Moscow Conservatory for studies with Heinrich Neuhaus. He developed into something of a maverick – an introverted man who had an unusual technique, an unusual repertory and an unusual manner of interpretation. His tempos could be unorthodox (often very slow) and his musical ideas different from those of all other pianists. On the stage he was, and continues to be, magical, taking his listeners along with him because of the obvious fervour of his beliefs and his brilliant way of giving them life. As Vladimir Ashkenazy has noted, everybody is swept away at a Richter recital. It is not until the event is over that questions arise. Richter seems to have examined the entire repertory, and he plays everything from Bach and Handel to Debussy and the latest Russians. His concerto repertory is very large. It was characteristic that he chose the little-played Dvořák concerto for his American orchestral début, and played it in Dvořák's original rather than the widely used, more virtuoso arrangement by Vilém Kurz.

Another maverick was the Canadian Glenn Gould (1932–82), as famous for his personal eccentricities as for his pianism. Gould was a Bach specialist, born to play Bach: he had fingers capable of perfectly calibrated scales and chordal weightings, and was uniquely able to delineate the linear strands of Bach's music. His repertory started with Sweelinck and Elizabethan composers, followed by Bach, Mozart and a handful of Beethoven sonatas. Then there was a big jump to Berg, Hindemith, Schoenberg and a few other moderns. He disliked nearly all Romantic music. He wrote articles expressing his distaste for Mozart and all of late Beethoven except the *Grosse Fuge*. He played from an extremely low position, using flat fingers. His Bach had a kind of bracing rhythm and accentuation that was entirely his own, and he was so convincing that he single-handed forced a revaluation of the Goldberg Variations, the Partitas, the '48' and other Bach pieces. Whether or not Gould's playing took note of musicological factors, it had tremendous character and life. Gould, as he had promised early in his career, retired around the age of 30 to make records and appear on television shows. He has become a legend to younger pianists all over the world, as much for the character of his pianistic genius as for his defiance of the establishment. His way of playing Bach and even his mannerisms have been widely imitated, especially in the USSR.

As the 1980s approached, the conventions of romantic piano playing were virtually gone. Horowitz, Arrau, Shura Cherkassky (*b*1911) and Jorge Bolet (*b*1914) were among the survivors. Cherkassky, Russian-born, studied with Hofmann at the Curtis Institute in Philadelphia and later moved to London. His playing can be rhythmically eccentric, but his singing line and colouristic resource are very much that of the old school. Bolet, also a Curtis product, developed late. He always had a technique of the Lhévinne order, but it was not until rather late in life that his playing developed the refinement, inner resource and tonal subtlety that have made him famous, especially as an interpreter of Liszt. He is one of the few living pianists whose playing makes veterans think in terms of Hofmann, Godowsky and Lhévinne.

In the 1980s Horowitz, Arrau, Bolet and Cherkassky were considered representatives of an old age. The new pianists were trained differently and sounded different. The period after World War II saw the emergence of the literal-minded musician who felt that his task was to play the notes exactly as written and to submerge his individuality to the intention of the composer. Romantic musicians considered themselves re-creators and worked on the premise that it was their duty to refract the message of the composer through their own personality. The new literal school was much more objective and the practitioners looked upon themselves as reproducers rather than co-creators. Structural considerations took precedence over emotional considerations. One unhappy result was a uniformity of style that extended from the Juilliard School in New York to the Moscow Conservatory. A prevalent complaint was that it was impossible to tell one pianist from another. All the conservatories seem to be a conveyor belt for producing competition pianists. It has been noted many times that many competition winners seem to disappear overnight. On the other hand it must be said that most of the admired pianists of the postwar generation have at some time been competition winners: Van Cliburn, Vladimir Ashkenazy, Maurizio Pollini, Murray Perahia, Radu Lupu, Krysztian Zimmerman, Martha Argerich, Emanuel Ax, Horacio Gutierrez and Garrick Ohlsson, among others.

The pianists of the new international school who have perhaps attracted the most attention are Alfred Brendel (*b*1931), Vladimir Ashkenazy (*b*1937), Maurizio Pollini (*b*1942) and Murray Perahia (*b*1947). Pollini and Brendel have developed into cult figures. The former represents the modern style *in excelsis*. He is a perfect executant with a flawless technique; he has a big repertory that extends into such

41. Maurizio Pollini, 1981

avant-garde music as the Boulez Second Sonata and pieces by Stockhausen; he is completely self-effacing on stage; he shapes the music with authority, clarifying its structure; and he stands outside the music, apparently refusing to become emotionally involved. Many young musicians consider him the greatest living pianist; they respond to his musicianly but cool approach and stand in awe of the sheer perfection of his playing.

Brendel has a more restricted repertory, confined largely to Bach, Mozart, Beethoven, Schubert and, surprising for one with his orientation, a great deal of Liszt. He also has played some Schoenberg. The archetype of today's so-called 'intellectual' pianists, his admirers regard him at the very least as a latter-day combination of Serkin and Schnabel. He has his detractors too, who find his playing pedantic and unimaginative. Some dismiss his Liszt playing out of hand because of his rather hard tone and lack of a super-virtuoso technique. Brendel, however, is more interested in the idea than the execution, and some of his ideas have brought new light to Liszt. He has discarded the Romantic ideas of Liszt playing and instead tries to bring to the fore the composer's prophetic elements – his adventurous harmonies and, especially in the later works, the other-worldly elements. This is Liszt seen through a very intelligent pair of 20th-century eyes.

Perahia has attracted considerable attention because of the elegance of his playing, his sensitivity, his singing line and the shapely proportions of his interpretations. It is small-scale playing, perhaps deliberately as Perahia does not have a very large repertory and has confined himself largely to Mozart, Beethoven and a few Romantic works. Very much of the modern school, he keeps strict rhythm, uses very little tempo fluctuation and tends, like Pollini, to avoid any high emotional commitment.

Ashkenazy as a young man was a poetic pianist who avoided virtuoso stunts (though he had a brilliant technique, as his early recording of Liszt's *Feux follets* demonstrates). Interchange with Western musicians after he left the USSR brought him into the international style. Ashkenazy, who has never lost his pianistic polish, may have lost something of the poetry and urgency of his youth as the years have gone by; everything in his large repertory he plays cleanly, accurately and literally – no more and no less than is written in the score. In recent years he has been spending more and more time conducting.

The Romantic revival of the 1970s does not seem to have had much impact on today's performing practice. Every age makes music its own way, and in the 1980s most pianists, skilful and dedicated as they are, seem to have abdicated the role of creative artist. Playing the notes with integrity, exactly as written, even within the conventions of an older period (to the extent that we can understand past conventions) may not be the answer to a re-creation of the music. There is a risk that modern musical scholarship, in giving performers so much valuable information about performing practice of the past, may inhibit them by handing down rules to be followed blindly. Present-day pianists should not forget, as many seem to, that composers have always been much more interested in the *Affekt*, the emotional meaning of their music, than in its architecture.

CHAPTER FOUR

Piano Music

Before the mid-17th century composers made little stylistic distinction between one keyboard instrument and another, and players used whichever happened to be available or was best suited to the occasion. It was not until the latter half of the 18th century that a distinctive style for the piano began to appear: the century after the death of Johann Sebastian Bach saw a dramatic rise in the popularity and prestige of the piano, both as a household instrument and as the vehicle for some of Western music's most enduring masterpieces. Although the principal contributions were made by relatively few composers, virtually all those active before World War I wrote music for or with piano.

1. THE ADVENT OF THE PIANO

The dominance of the harpsichord was not broken overnight; indeed, not until the dawn of the 19th century did the newer instrument altogether vanquish its plectra-activated rival. As late as 1802, Beethoven's three keyboard sonatas of op.31, though clearly designated for the 'pianoforte' by their composer, were published in Nägeli's series Répertoire des clavecinistes. Conversely, in 1732 Lodovico Giustini had published sonatas designated specifically for the 'cimbalo di piano e forte'. Although it became evident shortly after J. S. Bach had played on Silbermann's improved models in 1747 that the future belonged ultimately to the piano, the two designs coexisted peacefully throughout the second half of the 18th century. In January 1777 Mozart composed the Concerto in E♭, K271, on commission for a French *claveciniste* (i.e. harpsichordist). He performed it himself on a 'wretched' fortepiano in Munich in October 1777; the following January his sister played it on a harpsichord in Salzburg. The differences between performances on these two opposed instruments were narrower than they might seem today. The early piano was housed in a frame largely identical to that of the harpsichord, with equally light stringing. The fortepiano offered new possibilities for gradations in volume, but its tone was still characterized by the rapid decay of the harpsichord's. In terms of sheer sound, a triple-strung French double from this period produced as much, if not more, volume than its double-strung rival.

Conservative French composers such as Armand-Louis Couperin (1727–89) and Jacques Duphly (1715–89) continued to cultivate a lavishly intricate style perfectly suited to the opulent double harpsichords made by the Flemish builder Taskin. In Italy, the birthplace of the piano, Platti, Galuppi and others wrote music equally suited to either harpsichord or piano. The same interchangeability – doubtless designed to encourage sales – prevailed among the Iberians (Soler, Seixas, Blasco de Nebra), the Germans and Bohemians (some in Germany or Austria, such as Neefe in Bonn or Kozeluch in Vienna; others abroad, such as Schober and Eckhardt in Paris or Hässler in Russia), and the English (Nares, Hook). Carl Philipp Emanuel Bach, arguably the greatest keyboard player and composer in the generation after his father's, expressed a preference in his *Versuch* of 1753 for the subtle gradations and *Bebung* (vibrato) of the clavichord over any of its more extrovert relatives. In spite of their general designation as 'Clavier-Sonaten', the series from the 1760s and 1770s (often characterized as 'leichte' or 'pour l'usage des dames') were probably intended primarily for this most private of instruments. Along with the sonatas of Scarlatti, whose distribution turns out to have been far wider than was once believed, they exercised a considerable influence on the early sonatas of Haydn, who admitted: 'Anyone who knows me very well must realize that I owe a great deal to Emanuel, that I understood and studied him diligently'. Beginning in 1780 with C. P. E. Bach's second collection of *Sonaten nebst einigen Rondos . . . für Kenner und Liebhaber*, the 'fortepiano' is specified, a designation that carried through to his sixth and final set in 1787. Their composer revelled most in the kinds of dramatic contrasts of range and register that the new instrument made possible. Simple dynamic contrasts, though not as concentrated, are already called for in the six 'cembalo' sonatas dedicated to the Duke of Württemberg and published in 1744; these, achieved by discreet changes in registration, are fully realizable only on a two-manual instrument. The more complex range of effects that saturates the 'Kenner und Liebhaber' series – encompassing *pp* to *ff* and numerous shades in between – was scarcely equalled before late Beethoven. They are best understood as a natural extension of the registration shifts from three decades earlier. Nevertheless, as late as 1788 C. P. E. Bach was able to compose a Double Concerto for harpsichord and fortepiano, WQ47, in which the writing for the solo instruments is essentially identical; the chief delight lies simply in the tonal contrasts between them.

The rapid, experiment-orientated evolution of keyboard instruments during this period was reflected in the musical styles that flourished. The inevitable breakdown in High Baroque continuity was not to be fully replaced by Classical phrase structure until the 1780s; hence composers embracing *Empfindsamkeit* had to content themselves with a series of small-scale dramatic effects whose overall impact was often less than the sum of its parts. A great many movements in C. P. E. Bach's output fulfilling the minimum requirements of sonata form are diluted by the remoteness of secondary modulations and a surfeit of thematic material; indeed, only a composer of his extraordinary inventiveness could maintain interest amid such stylistic upheaval. His older brother Wilhelm Friedemann, in some respects even more gifted than Emanuel, never took final leave of his father's style. In an eclectic production that included sonatas, fugues and polonaises (these last enjoyed a vogue in the 19th century), nowhere was the dilemma of composers after the mid-century portrayed more clearly. Their younger half-brother Johann Christian shunned the complexities of the north for the relaxed *galant* style acquired during his formative years in Italy. His two sets of keyboard sonatas, opp.5 and 17, are model specimens of music created for domestic consumption: facile (though not without occasional technical challenges), diatonic to a fault, and highly polished. Between J. C. and C. P. E. Bach, virtually all the ingredients necessary for Viennese Classicism were present. Mozart seems to have acknowledged this when, although it was scarcely noticed in London, he mourned the death of J. C. Bach in 1782. About C. P. E. Bach he is alleged to have said: 'He is the father, we are the children'. As late as 1809 Beethoven could write to Breitkopf & Härtel that 'I have only a few items from Emanuel Bach's keyboard works, yet some of them not only provide the real artist with great pleasure, but also serve as objects to be studied'.

2. THE CLASSICAL SONATA

Although the music of the sons of Bach is among the earliest to benefit from sympathetic performance on the fortepiano, it is doubtful that any of them ever enjoyed the opportunity of performing on instruments as reliable as those praised by Mozart when he visited Stein's workshop in 1777. Even more than the singing tone, the composer was impressed by the regularity and evenness of the action, with its deceptively simple escapement. Though eventually rendered obsolete by the steadily increasing size of concert halls throughout the 19th century, Stein's design was both perfectly engineered on its own terms and perfectly suited to the world that Mozart was about to enter. After exclaiming that K284/205*b* (with its surprisingly lengthy set of variations as a finale) 'sounds exquisite' on Stein's instrument, Mozart – further stimulated by the Mannheim style with its emphasis on contrast – set down in the next several weeks two sonatas (K309/284*b* and 311/284*c*) more dramatically expansive and brilliant than any of the half-dozen surviving examples composed previously. These were succeeded the following summer by the first of his two sonatas in the minor mode, K310/300*d*, a work of remarkable intensity and tautness. In the space of a few years, and in direct response to developments in instrument design, Mozart had succeeded in transforming the easy-going three-movement form inherited from J. C. Bach (whose sonatas he had arranged as keyboard concertos at the age of nine) into a vehicle for considerable display and elaborate working-out.

With his final break from the archbishop in May 1781 and the decision to take up permanent residence in Vienna, Mozart inaugurated a series of masterpieces for keyboard dominated by 17 remarkable concertos, in which virtuosity is blended with a superb sense of operatic pacing. Though fewer in number, the ten solo sonatas now known to have been created after the move to Vienna (portions of K330–32/300*h*, *i*, *k* may have been composed a few months earlier) afford a unified view of the composer's development. A few, such as the 'little keyboard sonata for beginners', K545, were designed to fulfil pedagogic needs, but the remainder encompass a broad spectrum of mature styles. The group of four sonatas K330–33/300*h*, *i*, *k*, 315*c* (traditionally ascribed to Mozart's Paris sojourn of 1778, but now known to date from between 1781 and 1784) demonstrate his sure handling of practically every Classical form: sonata, both with coda (K332 finale) and without (K333 first movement); theme and variations (K331 opening movement); binary (K331 Menuetto and Trio); ternary (K330 Andante); rondo-type (K331 finale) and sonata-rondo (K333 finale). The last-named of these, with its tutti–solo opposition and elaborate cadenzas, offers a prime example of cross-fertilization with the concertos Mozart was composing during the same period. His treatment of all these forms is rarely perfunctory; the coda to the finale of K332 incorporates a *buffa* theme presented in the exposition but slyly omitted from the recapitulation. The 'Alla turca' of K331 adopts the thematic virtues of the straight rondo while employing an ingenious *ABCBAB* scheme to skirt its inherent structural squareness. The highly decorated version of the Adagio of K332 published by Artaria in 1784 (and presumably originating with Mozart) shows that improvised embellishment remained an integral component of his style; present-day performers might do well to contemplate the gulf between their abilities and Mozart's before undertaking their own decorations. The two-piano sonata, K448/375*a*, composed less than ten months after his arrival in Vienna on a commission from his talented pupil Josepha von Auernhammer, gravitates towards virtuoso display while displaying Mozart's intuitive understanding of the 'orchestral' capabilities of two fortepianos; the syncopated chordal responses in the opening Allegro's closing group are particularly striking. The composer's contact with the music of J. S. Bach and Handel at the concerts of Baron Gottfried van Swieten in 1782–3 resulted in a modest burst of contrapuntal works, including the underrated Prelude and Fugue in C, K394/383*a*, written at the urging of Constanze Weber.

Although Mozart soon tired of aping an archaic Baroque style, the effects on his own music of his experiences with Bach and Handel were profound and long-lasting. The unique single-voiced opening of K533 invokes the atmosphere of fugue, realized more fully in the second group, as well as in the minor-mode episode of the Rondo (published in 1788 with the two movements of K533 though composed in 1786). The opening movement of the Sonata in D K576, perhaps Mozart's masterpiece in this genre, bristles with lean, athletic counterpoint; it maintains the composer's predilection for the open-ended half-cadence that moves to the dominant in the exposition, while remaining in the tonic for the recapitulation (nearly half of the 35 major-mode sonata movements in the keyboard sonatas use this 'bifocal close'). Baron van Swieten's advocacy of C. P. E. Bach immediately stimulated two fantasias, K396/385*f* and 397/385*g*, both remaining fragments, although the second, in D minor, is still a favourite. The Fantasia in C minor K475, a work of great emotional scope, was published at the head of the sonata in the same key, completed five months earlier. Its impact on Beethoven's obsessive bouts with C minor can scarcely be exaggerated. A late Fantasia in F minor K608, composed in March 1791 for a mechanical organ but published as early as 1799 for piano four-hands, deserves more frequent hearings. Yet by far the most important development during this period was Mozart's deepening relationship with Haydn, whom he probably first met in 1781. Although Haydn's musical influence is most readily traceable in Mozart's mature chamber music, it is still felt in movements like the monothematic opening Allegro of K570, or in the bold choice of the lowered submediant as the secondary key of the Adagio of K576. The remarkable two-year period framed by the composition of *Le nozze di Figaro* and of *Don Giovanni* saw Mozart add four jewels to the crown of his works for keyboard, including the four-hand sonata K497, an unqualified masterpiece; an inspired set of four-hand variations K501; the chromatically rich A minor Rondo K511 and an outstandingly expressive Adagio in B minor K540. All this music was written for a five-octave instrument about which Mozart is not known to have voiced reservations. When the recapitulation of a sonata movement threatened to exceed its compass, his imagination was fired by the limitation, resulting in some of his most adroit touches, as in the opening movements of K333/315*c* (ex.6) or the

Ex.6 Mozart: K333/315*c*

concerto K449. The concert instrument used by Mozart and built by Anton Walter around 1780 included only two tone-modifying devices: a pair of knee levers that raised either all the dampers or only the treble ones (the presence of hand stops as well for the dampers on the original suggests Mozart may have requested the addition of knee levers, perhaps taking his cue from Stein's instruments); and a hand stop over the middle of the keyboard that placed a thin strip of cloth between the hammer and the strings, acting as a mute. In passages such as the middle section of the Andante of K330/300*h*, this *sourdine* imparts an ethereal effect fundamentally different from that achieved with the shift on a modern instrument. Both the mute and the raising of the dampers were regarded in Mozart's time as special effects; his

Ex.7 Mozart: K332/300*k*

celebrated remark that phrases should 'flow like oil' has often been construed as an unqualified endorsement of legato, inviting indiscriminate application of the modern damper pedal. In practice, both the rapid tonal decay on the fortepiano and the articulative richness of Mozart's scores preclude any uniform solutions. It is no condemnation of present-day instruments that the carefully marked phrasing at the opening of K332/300*k* (ex.7) is almost impossible to achieve naturally except on a fortepiano.

Haydn's reputation rested far less than Mozart's on his abilities as a keyboard performer. His longstanding positions as composer-in-residence to aristocratic patrons, including three decades of service to the Esterházy family, filled his days with the closely monitored composition of sacred, operatic, orchestral and chamber music, as well as with supervising performances. It is all the more surprising that Haydn found the time to compose over 60 multi-movement works for solo keyboard. Fewer than 50 of these can be proved authentic, and about a dozen more early harpsichord works were attributed to Haydn during his lifetime. As fewer than a dozen autographs (some only fragments) of Haydn's solo sonatas survive, the severe problems of chronology and authenticity among works circulating in the 1750s and 1760s are likely to remain unresolved unless new evidence is discovered. Most of these early pieces appear to have been teaching aids intended for the amateur, perhaps the children of Haydn's aristocratic patrons. It is unlikely that all, or even most, of them have survived. Entitled 'divertimento' or 'partita', they typically consisted of three movements, most often two fast outer ones encasing a minuet, though not infrequently with the latter as a finale. Apart from a few simple binary forms in works of questionable authenticity

(HXVI:7–9), virtually all the non-minuet movements present rudimentary sonata forms with modest transitions and well-demarcated secondary groups. Clearly designated for harpsichord, they exude the easy-going *galant* manner of Wagenseil without an obsessive reliance on the broken-chord basses purportedly popularized by Alberti. Significant increases in technical demands, perhaps stimulated by Scarlatti, are registered in the group of sonatas that includes HXVI:45, 19 and 46, composed in the late 1760s. The last movement of the Sonata in A♭ (no.46) foreshadows the irresistible *buffa* finales that Haydn was to perfect in the sonatas, quartets and symphonies of the 1780s and 1790s. Beginning around 1771 with the first works called 'sonate' (HXVI:18, 20 and 44), Haydn's unpretentious style is blended with increasingly complex emotional moods, easily traceable to the influence of C. P. E. Bach. The single dynamic marking in the autograph fragment of the Sonata in C minor (no.20) can still be rendered on a two-manual harpsichord, but by the time Artaria published this landmark in 1780 it included a wealth of additional dynamics (including a *crescendo* in the finale) that demanded the new flexibility of the fortepiano. The five other sonatas that appeared simultaneously (HXVI:35–9) are the last Haydn approved 'per il clavicembalo, o forte piano'. It may have been more than a coincidence that the trio of sonatas published in 1784 by Bossler (HXVI:40–42), and calling specifically for fortepiano, were the first that Haydn composed after the start of his friendship with Mozart. In 1788 Haydn wrote to his publisher Artaria that he had been compelled to purchase a new fortepiano in order to do justice to the three piano trios HXV:11–13.

Haydn's long life allowed him to continue to absorb and recast the most important advances of Viennese Classicism. The sonatas of Haydn's maturity are all the more remarkable for the stylistic distance that their composer had traversed to create them. The obligatory *da capo* minuet of previous decades disappears almost entirely; when required to supply one around 1789, the composer responded in the Sonata in E♭ (HXVI:49) with a large-scale 'Tempo di minuet' containing an elaborately rewritten repeat. A standard three-movement, fast–slow–fast scheme avoids tedium by incorporating at least one movement not in regular sonata form: the alternating major–minor variations (a favourite technique) that open HXVI:39 and close no.34; the spacious binary form with rondo elements that concludes HXVI:50; or the unexpected sonata-rondo that opens HXVI:51. But Haydn proved equally drawn in this period to a two-movement grouping, providing Beethoven with a point of departure for his subsequent experiments. Two of the three two-movement sonatas that appeared together in 1784 (in G and D) go so far as to abandon any references to sonata style. In the finale of no.40 Haydn took special delight in punctuating cadences with abrupt leaps of three octaves

Ex.8 Haydn: HXVI:40

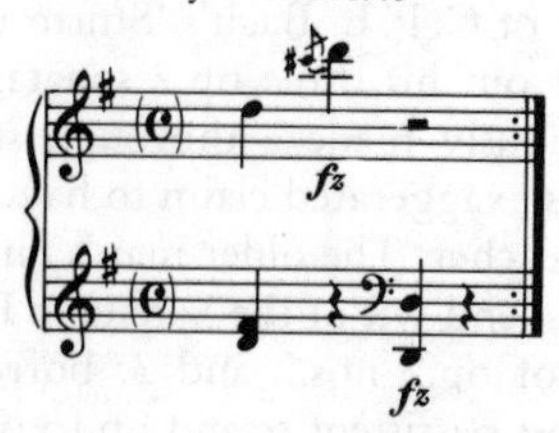

(ex.8); the fortepiano, with its clearly delineated registers, conveys the humour of these gestures with particular effectiveness. The pervasive imitation throughout the finale of the Sonata in D may reflect Haydn's encounters with J. S. Bach at Baron van Swieten's. Equally important is the surge of *cantabile* writing found in the slow movement of the Sonata in E♭ written about 1789 for Marianne von Genzinger, to whom Haydn extolled the virtues of a fortepiano by Wenzel Schantz. In the freewheeling Fantasia in C (HXVII:4), published around the same time, Haydn instructs the performer at two points to hold the cadential octave until the tone dies away; on a well-regulated modern grand the sound lingers for almost a minute. Between his first and second London sojourns, the composer penned an elaborate keyboard farewell to the double variation (HXVII:7), built on a pair of utterly non-symmetrical themes that erupt during only the third variation into a rhapsodic coda. Three highly individual sonatas (nos.50–52) composed during the next year in London provide a fitting climax to Haydn's output in this medium. The 'open pedal' demanded in the first movement of no.50 marks the migration of the Viennese knee levers to a location on the forward supporting legs of English models where they could be depressed with the foot. The finale of the same work exploits the $5\frac{1}{2}$-octave range of the newest English models; their fuller, weightier sound may be partly responsible for the symphonic grandeur that permeates the opening movement of no.52. Throughout his career Haydn's approach to sonata form was punctuated by surprise and experiment, continually nourished by his long-standing fascination with monothematicism. Even more than in the music of Mozart, Haydn's frequent changes of texture and spiky rhythms depend upon the quick response and rapid tonal decay of the early piano.

The most remarkable aspect of Beethoven's monumental 32 keyboard sonatas (including three teaching pieces in the spirit of Mozart's K545) is that they continue to expand and refine a genre that seemed to have reached perfection in the music of Haydn and Mozart. Three early sonatas (WoO 47) published before the composer was 13 present rather stiff imitations of C. P. E. Bach's 'Sturm und Drang' style. By the time he brought out his three op.2 sonatas in Vienna in 1796, Beethoven had obviously made a thorough study of Mozart and Haydn, in spite of his exaggerated claim to have learnt nothing from his most celebrated teacher. The older man's influence is easily traceable in the conciseness and wit of the Sonata in F op.10 no.2 or in the humorous scherzos of op.2 nos.2 and 3, borrowed from Haydn's quartets. But the most persistent strand up to op.22 is the loose, additive post-Classical language already discernible in Mozart's late piano concertos. Virtually every gambit in the opening movement of Mozart's K467 – the *piano* opening and subsequent tutti explosion, the bifocal close preceding a dramatic interjection of the minor dominant, the wealth of closing ideas that confirm the major – appear in the first movement of op.2 no.3, in the same key. The *con gran espressione* of op.7 and the Largo e mesto of op.10 no.3 invest Beethoven's slow movements with new dignity and pathos. Blatant sectionalism pervades the 'Rondo' finales of opp.7 and 22; here, as elsewhere, what separates Beethoven from the transitional generation of Clementi, Dussek, Hummel and Weber is his unflagging reliance on the sonata principle. By the 1790s the pressures on composers to abandon the symmetrical resolution of sonata form were considerable. Muzio Clementi, essentially a contemporary of Mozart who lived well into the new century, played an important role in these developments. His nearly six dozen keyboard sonatas published between 1779 and 1821 take Mozart as their point of departure (opp.7, 9 and 10 were published in Vienna), with greater emphasis on virtuoso techniques (such as the rapid parallel 3rds and octaves of op.2 no.4) and Italianate melody, especially in slow movements. After their contest before Joseph II on Christmas Eve 1781, Mozart characterized Clementi as a 'mere mechanicus'. The substantial increase over the next decade in the scale of his works is not matched by a corresponding increase in the capacity of thematic material to support the larger structures. Clementi's recapitulations frequently exhibit only a casual relationship to his expositions, with minimal attention paid to resolving long-range harmonic tension. The virtues of his last and best-known sonata, op.50 no.3, sub-titled 'Didone abbandonata', remain those of lean, athletic textures and dramatic changes of mood familiar from his earliest works. Curiously, although he was closely tied to piano manufacture from the 1790s, little of the increased capacity of the new six-octave instruments is reflected in Clementi's keyboard music, probably because most of it was composed by 1805.

Between 1817 and 1826 Clementi brought out a series of volumes under the title *Gradus ad Parnassum*, devoted to the attainment of a fluent technique. Debussy paid an affectionate tribute to the popularity of these exercises in his 'Doctor Gradus ad Parnassum' from *Children's Corner*. Clementi was joined in these endeavours by two other distinguished men, Carl Czerny and Johann Baptist Cramer. Czerny had studied as a youngster with Beethoven before becoming a private instructor from the age of 15, numbering among his pupils Kullak, Thalberg, Heller and the young Liszt. Although Liszt frequently played Czerny's Sonata no.1 in A♭ op.7, it was as an indefatigable pedagogue that Czerny chose to make his mark. In more than 800 works devoted largely to technical studies (the best known being the *Complete Theoretical and Practical Pianoforte School* op.500), Czerny compiled and codified the technical advances of the piano during a period of extremely rapid development. If Czerny's methods were already beginning to show signs of age before his death, he continued to command the respect and admiration of his peers. Cramer, although an essentially conservative force like Czerny, was (according to Ferdinand Ries) considered by Beethoven to be the finest pianist of his day. He is remembered chiefly today for two fine sets each of 42 studies, published in 1804 and 1810 and endorsed by Beethoven, Schumann and Chopin.

Foreshadowings of at least a dozen composers from Beethoven and Schubert to Liszt and Brahms have been detected by proponents of the music of Dussek. In terms of pianistic figuration, there is no doubt that Dussek was a pioneer; formally he was much less so, relying heavily on the rondo and other sectional schemes. No hard evidence remains to show that Beethoven was familiar with his music, as can be demonstrated in the case of Clementi. Nearly 30 sonatas (several bearing programmatic titles) composed between 1788 and 1812 bear witness to a highly eclectic style stimulated by Dussek's peripatetic career as a travelling virtuoso. His association with the firm of Broadwood contributed to an expansion of the piano's range to six octaves (*C′–c′′′′*) as early as 1794. Hummel's ties to Viennese Classicism were considerably stronger, for he had studied with Mozart as a child and returned frequently to Vienna. Until the 1820s Hummel's fame nearly rivalled Beethoven's. Apart from an early sonata issued in London, his five remaining works in this genre were published in Vienna between 1805

Ex.9 Hummel: op.81

and 1825, including a near-masterpiece, the Sonata in F♯ minor op.81, which appeared just after Beethoven's op.106. The exposition of its opening movement arrives in A major after a generous interlude in C major, pointing up Hummel's continued loosening of high Classical structures, as well as his anticipation of Schumann's harmonic palette (ex.9). Like Clementi's and Dussek's, Weber's career was marked by extensive travels; unlike either, his principal field of activity was opera. When, on examining the score of *Der Freischütz* in 1823, Beethoven remarked that its composer 'must write operas, nothing but operas', he displayed a keen appreciation of Weber's special gifts. Throughout his four sonatas (all but the third in four movements) the pacing is consistently operatic, aided by directives such as *con duolo*, *mormorando* and *consolante* in no.4. Running passage-work over simple chordal accompaniments, as in the first movement of the Sonata in A♭, look forward to such patterns in the works of Chopin. For his own part, Weber remarked in 1810 that Beethoven's compositions after 1800 were 'a confused chaos, an unintelligible struggle after novelty'.

Weber was almost certainly referring to Beethoven's resolve not to settle into the structurally less demanding language of the proto-Romantics. In the highly experimental sonatas of opp.26–8 it looked as if Beethoven might indeed pursue this path. The A♭ Sonata dispenses altogether with straight sonata form. Both of the op.27 sonatas exhibit novel structures, and op.28 is noteworthy for its off-tonic beginning and third-related modulatory scheme. The conflicts in Beethoven's style around 1800 are drawn cleanly in op.27 no.2 (the 'Moonlight'), whose famous opening demands the intimacy of the drawing-room, while its stormy and very public finale pushes the five-octave instrument inherited from Mozart right to (though not beyond) its limits. Op.31 no.3 was the last four-movement sonata until the inaptly labelled 'Hammerklavier' (the generic term for the Viennese piano after 1815) of 15 years later. In the autumn of 1802 Beethoven wrote to the publisher Breitkopf & Härtel concerning the 'new manner' of his two sets of variations, opp.34 and 35. Continuing with the 'Waldstein', and even more emphatically with the 'Appassionata', Beethoven re-created the taut, integrated aesthetic of the high Classical period, though on a greatly intensified scale. It scarcely seems an accident that this dramatic turnabout in Beethoven's style paralleled equally dramatic developments in the Viennese piano. Within six years the instrument nearly doubled its weight and more than trebled its string tension. The menacing opening of op.57, plumbing the lowest note on the keyboard, is unthinkable without the powerful yet clear bass of the new six-octave models. The lush sweetness of these instruments is reflected in the two movements of op.78, Beethoven's only work in F♯ and a particular favourite of the composer's. 'Les Adieux', op.81*a*, composed in the same year and key as the 'Emperor' Concerto, provided a fitting close for the solo sonata to the 'heroic decade'. Both opp.90 and 101 show a closer affinity with the later styles of Schubert and Mendelssohn respectively, revealing a composer once again at the crossroads. Much like op.57 of a dozen years earlier, the monumental Sonata in B♭ op.106 marked Beethoven's final return to an expanded vision of the high Classical style, spurred by another burst in the size and weight of Viennese pianos. The frequent choice of non-dominant secondary areas in sonata movements after 1817 is overshadowed by continually deepening levels of thematic integration, such as the relentless chains of descending 3rds that saturate the first movement of op.106 (ex.10). The Adagio of this remarkable work, placed after the Scherzo and in the remote key of F♯ minor, is both the longest and the most deeply felt among Beethoven's slow movements. But it was the composer's renewed interest in fugue, first seen in the finales of op.101 and the cello sonata op.102 no.2, that dominated the late style. The equally fugal yet diametrically opposed finales of both opp.106 and 110 demonstrate the extent to which Beethoven could impose his will upon the intractable rules of counterpoint. Closely allied with this absorption was the practice of variation, culminating in the Arietta of op.111, whose transcendent blend of variation and sonata inspired Kretschmar's impassioned homage in Thomas Mann's *Dr Faustus*. When invited to contribute a variation on the publisher Diabelli's 'Schusterfleck' of a waltz, Beethoven responded over a period extending from 1818 to 1823 with a series of 33 variations that constitute a final compendium of Classical techniques. He took his leave from the piano with his third cycle of (as Beethoven referred to them) Bagatelles op.126, which not only served as an experimental laboratory for the late quartets but also anticipated the character-pieces of the Romantics.

Although Schubert never billed himself as a pianist, he produced a prodigious quantity of keyboard music over scarcely more than a decade, including 11 solo sonatas, substantial fragments of nine others, three sets of Impromptus and *Moments musicaux*, and more than 400 dances for occasional use. During his lifetime the 16-bar *Trauerwalzer* D365 no.2 became so popular that its citation did not require the identification of Schubert as the composer. He began half a dozen sonatas before completing D537, the first of three impassioned works in A minor. Two of these, along with the 'little' A major (a perennial

Ex.10 Beethoven: op.106
ff
fp
f
fp
p
sempre
sempre p
cresc.
più cresc.
f

favourite) are in only three movements; otherwise Schubert – unlike Beethoven after 1802 – preferred the spaciousness of a four-movement plan. Among the dance movements scherzos are most represented, but a work as late as the Fantasia in G D894 (1826) presents an old-fashioned Menuetto. In certain respects Schubert was formally less experimental than Beethoven. All of his opening movements are in sonata form; after 1819 all but one of his finales is a sonata-rondo or an even simpler straight rondo. His slow movements are slightly more adventurous, favouring the two- and three-part forms whose simple contrasts proved so appealing to the next generation. But it is the relationship in Schubert's music between theme and tonality that differentiates him from his great contemporary and that so profoundly influenced Brahms and Mahler. The 'heavenly length' praised by Schumann points up the leisurely unfolding of long, arching themes rooted in song. Rather than struggling to create dynamic transitions along Beethovenian lines, Schubert viewed the obligatory modulation in expositions as an opportunity for a series of bold, common-tone key changes that minimize the structural significance of the secondary tonality. In movements like the finale of the C minor Sonata D958 this process is carried to almost bizarre lengths; in others, such as the deeply moving Molto moderato that opens the last of the late sonatas, D960, the motion through the flattened submediant (both major and minor) is achieved effortlessly through what amounts to thematic transformation. Schubert's models in these sonatas, which compare in importance with the late sonatas of Beethoven, are clearly the mature sonatas of Hummel (to whom he planned to dedicate his final three). Although lacking the technical challenges routinely confronted in Beethoven's music, their figuration is rarely perfunctory; a compelling performance demands an outstanding sensitivity to proportion and pacing. The two exceptions to these moderate technical demands are the Sonata in D D850, composed during the same summer, that of 1825, which saw the composition of the 'Great' C major Symphony, and, emphatically, the *Wanderer* Fantasy, a work of unabashed virtuosity whose continuous structure inspired the cyclic forms of Liszt. The song that provides the starting-point for its slow section, and from which the work derives its name, provides perhaps the most splendid example in Schubert of the poignant contrast between major and minor.

Schubert's interest in smaller forms ran considerably deeper than Beethoven's, and resulted in some of his finest efforts. The two sets of four impromptus and the six *Moments musicaux* (a title invented by the publisher Leidesdorf) were created largely in the last two years of the composer's life, at least partly in response to exhortations from publishers for less demanding music. It is a tribute to Schubert's greatness that he was able to produce masterpiece after masterpiece among works directed solely at the domestic market. Only the first of the op.142 impromptus uses sonata form, inspiring some writers to interpret its other three members as the remainder of a four-movement sonata. At least half of the 14 pieces in these works are straightforward ternary forms with verbatim repeats of their opening sections. Others, such as op.94 no.2, introduce the double variation (*ABA'B'A"*) inherited from Hadyn and later exploited by Mahler. The care lavished by Schubert on the countless sets of *ländler*, German dances, waltzes and ecossaises (the first three of these stylistically indistinguishable) far exceeded the demands of the form; many invite enrichment by the discreet addition of the pedal-activated buff or Janissary stops in vogue during the first quarter of the 19th century. Their application was mandatory in the fashionable battle pieces first popularized by Kocžwara's *The Battle of Prague* (*c*1788). Although Schubert rarely exploited the available range of the Viennese pianos (none of the last three sonatas uses the extra 4th added in the bass around 1816), his relationship to these instruments is considerably more sensual than that of Beethoven. The idiosyncratic wide spacing of chords, so frequently featuring the 3rd in the soprano, and the placement of tunes in the clear, singing tenor register reflect the special virtues of the pianos on which Schubert composed and performed.

Schubert's achievements in smaller forms were not without precedent in works by two Bohemian composers, Jan Tomášek and Jan Voříšek. With a series of evocatively titled eclogues, rhapsodies or dithyrambs published between 1807 and 1818, Tomášek laid good claim to being the originator of the short character-piece that proved so appealing to Romantic composers. His pupil Voříšek took up residence in Vienna, where he enjoyed fruitful relationships with Beethoven, Hummel and Schubert. Although documentation is lacking, it seems likely that Voříšek's impromptus influenced Schubert's compositions of the same name.

3. ROMANTICISM AND THE MINIATURE

After the deaths of Beethoven (1827) and Schubert (1828) the decline of the sonata was swift and precipitous. Although its prestige remained enormous, largely because of the achievement of

Beethoven, stylistic developments turned rapidly in other directions. The sonatas of Schumann, Chopin and Brahms, however imaginative in certain respects, project a sense of imitation rather than continued evolution. In Germany the chief architect of this aesthetic shift was Robert Schumann, who used his editorship of the *Neue Zeitschrift für Musik* as a forum for proclaiming both Chopin and the young Brahms. Composed during the 1830s, Schumann's first 23 opus numbers were all for solo keyboard, including several of his best-known works. From his op.1 (the 'Abegg' Variations) on, the voice is clear and assured, characterized by an extraordinarily poetic harmonic imagination, strong root movements, frequent doublings and a preference for the middle range of the piano. Although Robert and Clara did not receive the grand manufactured by Conrad Graf until their marriage in 1840 (the instrument was later bequeathed to Brahms), the music composed by both demonstrates the warmth and intimacy of the Viennese instruments. Many of his most successful works, including *Papillons* op.2, the *Davidsbündlertänze* op.6, *Carnaval* op.9 and *Kreisleriana* op.16, consist of cycles of miniatures whose interdependency is analogous to that found in the later song cycles like *Dichterliebe*. In *Carnaval* a series of epigrammatic mottoes provides a modicum of musical connection, but the deeper unity is more elusive, based on harmonically open beginnings or closes and a keen sensitivity to contrasts in mood. Along with figures from the *commedia dell'arte*, Schumann presents sympathetic portraits of Clara Wieck, Chopin and Paganini, as well as of Eusebius and Florestan, the introvert and extrovert sides of his own musical personality. It is surprising to find the density of short internal repeats – betraying binary origins – in movements of wide-ranging harmonic freedom like those in *Kreisleriana* (inspired by E. T. A. Hoffmann's character and dedicated to Chopin). Often accompanying these repetitive forms are the kinds of motoric rhythm familiar from the Baroque (Schumann acknowledged that his music was closer in spirit to Bach than to Mozart). The predilection for building on short, symmetrical harmonic sequences can lead to a marked squareness, often rescued by highly original figuration. Apart from the opening movements of the three sonatas, sonata form surfaces only rarely in Schumann's works. An effective example is the finale of the *Faschingsschwank aus Wien* op.26, whose opening rondo remains one of the composer's freshest inspirations.

In such works as the *Studien nach Capricen von Paganini* op.3, the Toccata op.7, the *Symphonische Etüden* op.13 and the *Phantasie* op.17, Schumann made important contributions to an expansion of the piano's range and sonority, keeping pace with the new iron-framed instruments being built in the 1830s. The *Phantasie*, dedicated to Liszt and whose proceeds Schumann contributed to the fund for the Beethoven monument in Bonn, is considered by many to be his masterpiece. With its pointed references to the last of Beethoven's songs from *An die ferne Geliebte*, it offers an eloquent farewell to Classicism. In spite of a reliance on structures of the *da capo* type and strong subdominant leanings, it is one of Schumann's most successful large-scale works, concluding with a serene slow movement in C that evokes the spirit of the Arietta finale of Beethoven's op.111. Schumann's considerable reliance on the metronome has been attacked on numerous occasions, but used with care (and sometimes modified by Clara's own editorial suggestions) his markings provide a very useful guide. He was also one of the first composers to designate long passages as simply 'mit Pedal', confirming the shift of the dampers' function from that of a special effect to a continuous ingredient in the texture. Finally, Schumann's commitment to high-quality pieces in his studies for children resulted in such welcome additions to this repertory as the *Kinderszenen* op.15 and most especially the *Album für die Jugend* op.68.

Although Schumann's innovations appeared less radical by the end of the century, they remained more far-reaching than those of his contemporary Mendelssohn. After leading a revival of the *St Matthew Passion* in 1829, Mendelssohn issued a series of keyboard works that included preludes and fugues (a set of six appeared in 1837), capriccios and fantasias, evoking a Baroque atmosphere overlaid with post-Classical phrase structure. A favourite arrangement was the slow introductory opening succeeded by a fleet Allegro or even Presto, most familiar from the *Rondo capriccioso* op.14, composed when Mendelssohn was only 15. A quarter of his output consists of eight books of *Lieder ohne Worte*, shorter lyric pieces predominantly in simple ternary form, whose moderate technical demands offered sustenance to the amateur player in danger of being swamped in a sea of virtuosity.

The designation 'revolutionary' is properly reserved in the 19th century for a figure like Chopin. In spite of precedents to be found in the music of Hummel and Field, even Chopin's earliest works are stamped with an originality that could scarcely have been expected. All of his more than 200 works involve the piano (the vast majority are for piano solo), and in this respect he typifies the increasing specialization of the Romantics. Only a handful of concertos, sonatas and chamber works employed what were by now academic forms. Otherwise Chopin preferred generic titles that readily conjured up poetic images (ballade, barcarolle), though he stopped short of overt programmaticism, maintaining the tradition of absolute music in the two composers he most revered, Bach and Mozart. His discomfort with large, multi-movement forms is betrayed in the two youthful concertos, whose opening movements reverse the customary sequence of modulations in exposition and recapitulation. A visit to Vienna in 1829 saw the première, on an instrument placed at Chopin's disposal by Conrad Graf, of the variations on 'Là ci darem la mano' (the work to be greeted by Schumann's prophetic review: 'Hats off, gentlemen, a genius!'). The Waltz in E♭ op.18, the first of the large concert waltzes, was also set down in Vienna. But Chopin's decision in the autumn of 1831 to take up at least temporary residence in Paris sealed the decline of the imperial capital and marked the ascendancy of the French metropolis to its position as the centre of new musical fashion for the next 90 years. Most importantly for the evolution of the piano, developments now shifted to the French–English design. Both the more conservative English action retained by Pleyel and the double escapement patented by Erard in the 1820s (the model for virtually all modern grand actions) provided more leverage with less effort than the increasingly cumbersome Viennese action, whose mechanical disadvantage multiplied as the instruments grew in size and weight. Many of Chopin's effects depend upon the increased sustaining power, particularly in the treble, of the newest French instruments. At the same time, both Pleyel and Erard's flat-strung pianos retained a clarity and transparency, even in the bass, that was aided by a more lightweight and efficient damping system. Gone for good were the exotic multiple pedal stops of the Viennese instruments; Romantic pianists made do with the damper and shift pedals now standard on English models. Chopin's preference for the more intimate sound of the Pleyel (whose action was slightly shallower than that of the Erard and had virtually no after-touch) shows that he resisted over-simplified notions of 'progress'.

On his arrival in Paris, Chopin began the regular and systematic cultivation of almost ten different genres. Dominant among the smaller forms were the mazurkas and nocturnes, which collectively reveal an astonishingly varied approach to ternary form. The modal colouring of the Mazurka in C♯ minor op.41 no.4 (caused by the use of the lowered 2nd and 7th degrees) sets up the return to the opening *A* section via the augmented 6th rather than the dominant, a technique that was to become a Romantic cliché. While still a youth in Warsaw Chopin had access to an intriguing new genre of composition by the Irishman John Field: the nocturne. The first four examples bearing this atmospheric title appeared in St Petersburg and Moscow in 1812, and doubtless made their way to Warsaw soon after. To Field goes the credit for evolving the arpeggiated accompaniment over which an expressive melody is free to spin out. Traces of the nocturne as it was inherited from Field are evident in op.9 no.2 (a perennial favourite of amateurs) but Chopin soon transformed the species to accommodate a much wider emotional range. The extreme contrasts of op.15 no.1 provide a memorable early example; 15 years later the highly ornamented return in the Nocturne in B op.62 no.1 raises subtlety to new heights while assimilating Chopin's love of Italian *bel canto*. Although performers frequently present them in different groups, many of these sets were arranged by Chopin as collections unified in sequence of mood and tonal plan. The almost 20 waltzes are more openly sectional, as befits their dance origins, and prompted some of the composer's most spontaneous melodies, reinforced in the larger concert waltzes by ingenious repetition schemes. His most direct homage to Bach, the 24 Preludes, encompasses an array of formal schemes far richer than their aphoristic character might suggest. A large number are built upon a single phrase that requires only a single repetition rather than contrasting material to attain completeness. An even more virtuoso treatment of repetition underlies the Berceuse op.57, where a simple alternating pattern of tonic and dominant harmonies repeated 54 consecutive times supports a remarkably free and florid set of seamless melodic variations. A similar union of circumscribed harmonies and operatic display (frequently in duet textures) informs the equally remarkable Barcarolle op.60, which captures perfectly the gentle undulations of a Venetian gondola without the sentimentality so often attached to the genre.

Apart from his one youthful sonata, Chopin's experiments in this form produced two highly individual works, both in the old-fashioned four movements though with the scherzo placed second. In both opening allegros the focus on thematic rather than tonal processes leads to a marked sectionalization between vigorous first and lyrical second groups. The finale of the B♭ Sonata is one of the most original movements Chopin ever wrote, subjugating all the traditional elements to a single, bare, fleeting texture. His ten or so remaining large-scale works (all in one movement) evince two opposed approaches. The polonaises, the first three of the scherzos and the second of the ballades employ large-scale ternary or rondo structures built around highly contrasted material. However, the three remaining ballades (in G minor, A♭ and F minor), as well as the Scherzo in E op.54, the Fantasie op.49 and the Polonaise-fantaisie op.61, each offer highly individual solutions to the special formal problems posed by thematic transformation and seamless transitions. The influence of sonata procedures is obvious in the first and last ballades and in the Fantasie, though with a minimum of emphasis on resolving material from secondary keys in the tonic. By establishing A♭ major as emphatically at the close as it does F minor in its opening, the Fantasie promotes the interchangeability of relative major and minor; the conclusion in A minor of the F major Ballade, which made such an impression upon Schumann, provides an even stronger example of Chopin's undermining of a single, central tonality. Although the Fantaisie-impromptu, published posthumously, has always been the most popular of Chopin's compositions in this vein, his 'fantasy' masterpiece is doubtless the Polonaise-fantaisie, in which the most heroic and extrovert characteristics of the genres cultivated by Chopin are blended with the most intimate flights of fancy. Performances that ignore the single basic tempo marking of Allegro maestoso obscure the underlying unity.

4. THE AGE OF VIRTUOSITY

Keyboard virtuosos had travelled across Europe since the mid-18th century, but the bulk of published music was aimed at the amateur market. Beginning with Beethoven, the situation was rapidly transformed; Czerny reported to the composer in a conversation book that a woman in Vienna could still not play the opening of the 'Hammerklavier' even though she had been practising it for months. The 84 studies of Cramer, published in 1804 and 1810, were considered by Beethoven to be the 'best preparation' for his own works, receiving praise in the next generation from Schumann. Czerny's *Complete Theoretical and Practical Pianoforte School* op.500, although not published until 1839, codified earlier practices. The era of the Romantic virtuoso was properly launched with the publication of Chopin's two sets of études in 1833 and 1837 (though the earliest were composed in 1829). He combined the solution to a single technical problem (including rapid parallel 3rds, 6ths, or octaves in the same hand; black keys, large jumps) with works of intrinsic artistic merit, worthy of placement alongside any others in the concert repertory. Schumann's description of op.25 no.1 as 'a lovely picture in a dream' acknowledges Chopin's highly original figuration, in which 'it would be a mistake to suppose that he allowed us to hear every one of its small notes'

Ex.11 Chopin: op.25 no.1

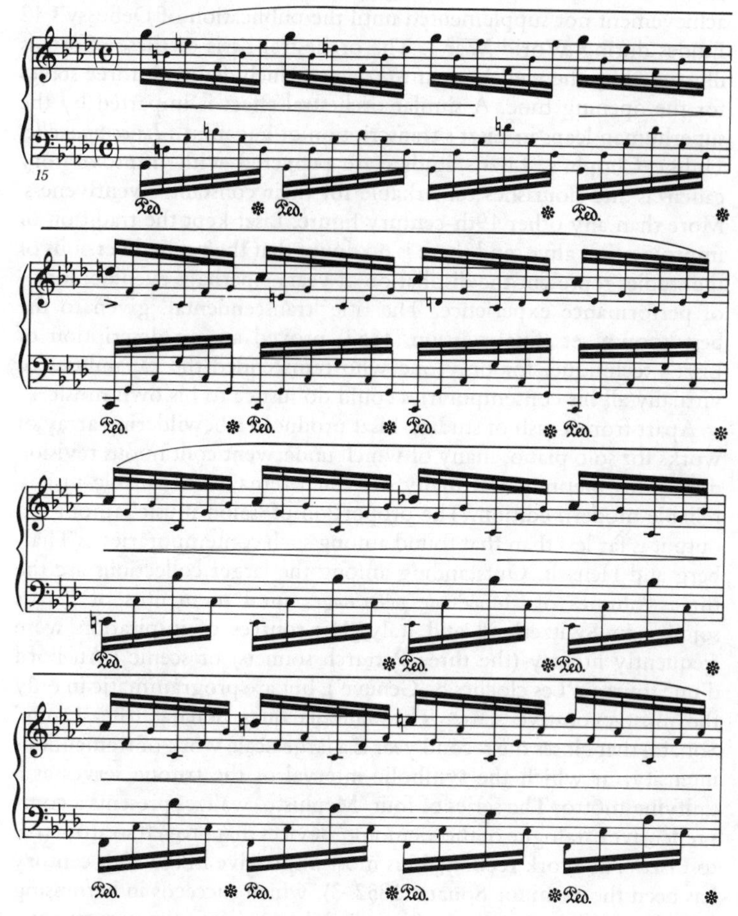

(ex.11). He was equally adroit in studies that develop touch rather than bravura, especially evident in the three composed in 1839 for inclusion in Moscheles's *Méthode des méthodes*.

The only 19th-century performer capable of doing justice to the expansive arpeggios of Chopin's op.10 no.1 was said to have been Franz Liszt, and it was he who carried the evolution of the Romantic pianist to its fever pitch. Beginning at the astonishingly early age of 15, and inspired by the example of Paganini, Liszt published between 1826 and 1849 (he retired from concert touring in 1848) almost three dozen studies encompassing a dazzling spectrum of keyboard effects, an achievement not supplemented until the publication of Debussy's 12 Etudes during World War I. The orchestral basis of these efforts is illustrated by the well-known *Mazeppa*, which demands three staves for the opening tune. A similar orchestral effect is imparted by the superhuman leaps in Liszt's transcription of Paganini's *La campanella*. Unlike Chopin's, Liszt's studies are peppered with improvisatory cadenzas and flourishes remarkable for their constant inventiveness. More than any other 19th-century figure, Liszt kept the tradition of improvisation alive, and there is no doubt that the printed versions of the studies represent the distillation of years – perhaps even decades – of performance experience. The title 'transcendental' given to the best-known set (final version, 1852) proved an apt description of Liszt's technique, for only one who transcended the capabilities of virtually all his contemporaries could do justice to his own music.

Apart from a rash of studies, Liszt produced a bewildering array of works for solo piano, many of which underwent continuous revision during his lifetime, and many of which remain unavailable in any reliable modern edition. The proportion of 'salon music' among his output is far less than that found among such contemporaries as Thalberg and Henselt. Outstanding among the larger collections are the three volumes of *Années de pèlerinage*, aural mementos of Liszt's sojourn in Switzerland and Italy. His sources of inspiration were frequently literary (the three Petrarch sonnets) or scenic ('Au bord d'une source', 'Les cloches de Genève'), but are programmatic in only the most evocative sense. The 'fantasia quasi sonata' (the 'Dante' Sonata) that closes the second year is a large-scale work of tremendous intensity, in which the symbolic interval of the tritone serves as a unifying motto. The series of four 'Mephisto' waltzes presents a comprehensive catalogue of the 'demonic' devices that proved so attractive to Liszt. The work reckoned his most impressive in the 20th century has been the B minor Sonata (1852–3), which succeeds in harnessing technical brilliance to the architectural demands of four-movements-in-one. The sonata is perhaps Liszt's most impressive display of thematic transformation, built upon an edifice of five mostly cryptic and open-ended motifs. It would be a serious error, however, to overlook the tremendous investment made by Liszt in arrangements, transcriptions and works based on previous material. Most important among the latter are the 21 Hungarian Rhapsodies based on processed folk material, planting the seeds for the nationalistic movements at the end of the century. Liszt's high opinion of Schubert is reflected in the more than 60 song transcriptions, including the complete *Schwanen-*

gesang and *Winterreise*. His many operatic transcriptions and paraphrases are now rarely heard, but in his own day they not only provided opportunity for technical display but served many of the functions of the gramophone. Liszt lavished considerable care upon such arrangements, and in his 'Reminiscences from Mozart's Don Juan' he left behind a graphic representation of technique as sexual conquest.

Although much has been made of Liszt's enthusiastic endorsement of Steinway's new overstrung models in the 1870s, the vast majority of his music for piano was composed during the period in which he endorsed the flat-strung Erards with equal enthusiasm. He even found time to provide testimonials for Chickering, and for the Bösendorfer with its old-style Viennese action. In any event, all the instruments used by Liszt were equipped with softer wire and more elastic accretions of felt and leather hammer coverings than 20th-century concert instruments. His long career spanned a phenomenal period in the piano's development, and he never tired of dreaming up new and seemingly unattainable effects, such as the 'vibrato assai' in his transcription of Schumann's *Widmung* (ex.12).

Liszt's achievements inspired both competitors and imitators. His

Ex.12 Liszt: Transcription of Schumann's *Widmung*

a tempo
Geist, mein bess - res Ich!
con somma passione
fff vibrato assai

cont. overleaf

sharpest competition in the late 1830s was from Thalberg, who dazzled audiences with his novel device of placing the melody in the thumbs while surrounding it with a sea of arpeggios, giving the impression that more than one piano was being played. Thalberg specialized in operatic paraphrases (that on Rossini's *Moïse* enjoyed particular popularity) and variations such as those on *God Save the King*; none of his extensive output remains in the active repertory today. A similar fate has befallen the transcriptions and salon pieces of two other celebrated virtuosos, Herz and Henselt. The most interesting and original pianistic figure next to Liszt in the mid-century was Alkan, who spent much of his life in obscurity. Novel (and sometimes epic) notions of structure and harmony have served to rekindle interest in Alkan's music, the variety of which rivals that of his better-known contemporaries. His virtuosity was uncompromising, at times requiring an almost superhuman stamina.

Brahms's virtuosity took Beethoven's 'Hammerklavier' as its starting-point, evidenced in the rhythms and proportions of his C major Sonata, published when he was scarcely 20. After the three early sonatas, however, Brahms turned his attentions elsewhere. The chief focus during the late 1850s and 1860s was variation form. The 25 Variations and Fugue on a Theme of Handel op.24 injected new life into a genre virtually moribund since Beethoven's set for Diabelli four decades earlier. Brahms summarized his technique – more severe and less effect-orientated than that of Liszt – in two striking sets of variations on Paganini's Caprice no.24. Typical among the uncompromising problems aired are the 'blind' octaves in no.11 of the second book (ex.13). Beginning with the Eight Piano Pieces of op.76, published when he was in his mid-40s, Brahms focussed almost exclusively for the next 15 years on six groups of smaller pieces described variously as Capriccio, Intermezzo, Rhapsody, Ballade or Romanze. Although he occasionally included literary inscriptions (from Sternau over the Andante of the F minor Sonata, from Herder at the beginning of op.117),

Brahms's fundamental allegiance remained with the abs tradition of the Viennese Classicists. Strife between him anc garde advocates of Liszt and Wagner proved inevitable. A fe shorter works fulfil the dramatic demands of sonata form (the Capriccio and the B♭ Intermezzo from op.76), but Brahms relie ost heavily, as had Chopin and Schumann before him, on the simple ternary scaffolding. If he rarely infused it with the endless flexibility of Chopin, Brahms's resourcefulness, particularly in matters of rhythm and phrase, rarely faltered. Regardless of mood, he gravitated towards the middle and lower registers of the piano, preferring chains of closely spaced, poignant dissonance to clearly articulated textures. In spite of opportunities to experiment with the newer, high-leverage actions, Brahms remained loyal until the very end to the Viennese models that soon after his death were to pass into obscurity. He remains one of the few composers in the Western tradition for whom nostalgia for a bygone era provided a fresh and original impulse.

5. 19TH-CENTURY NATIONAL TRENDS

By the 1870s the piano and its literature had attained a pre-eminence unrivalled both in the salons of the upwardly mobile middle class and on the concert stage. It claimed a repertory from Bach to Brahms that was, and remains, beyond comparison in its scope and its extent. To expect the flood of masterpieces that had issued forth for almost a century to continue indefinitely would have been unrealistic even had it not been that the piano's popularity reached a peak, to be followed by a shift of focus back to the orchestra. The piano continued to inspire composers and performers alike, but much of the activity now took place beyond the main arenas of Germany, Austria and France.

Ex.13 Brahms: op.35

In Mozart's time there had been relatively little distinction between teaching or domestic pieces (sonatas, variations) and those intended for public consumption (primarily concertos and chamber music). After Beethoven's death the emphasis among professionals on the development of a 'superhuman' technique (assisted by mechanical aids such as finger stretchers and dumb keyboards) led to a bifurcation of the solo repertory. A few major composers like Schumann attempted to fill the void with instructional cycles of high quality (*Album für die Jugend*); others such as Stephen Heller, who also composed large quantities of ambitious music, are remembered primarily for a steady stream of undemanding pieces aimed at the amateur market.

As in opera and orchestral music, nationalist piano music betrayed considerable western European influences, particularly that of Liszt. Almost all the Russians wrote for the piano. The salon pieces of Glinka, Borodin and Rimsky-Korsakov are surpassed in interest by those of Tchaikovsky, but it was two other Russians who made the major contributions. Perhaps the most original of these was Musorgsky's *Pictures at an Exhibition* (1874), a series of tableaux inspired by paintings of Victor Hartmann and linked by a recurring promenade theme in 5/4 metre. The writing, both stark and colourful, captures the folk flavour more effectively than Ravel's opulent orchestration. Balakirev's *Islamey* (two versions, 1869 and 1902) has acquired a certain status as the technically most demanding work in the virtuoso repertory – too difficult even for its composer, an accomplished pianist – but it is also skilfully written and dramatically effective.

The English-speaking world boasted its most successful 19th-century keyboard composer in Sterndale Bennett, most of whose music is unknown today. Admired by Schumann and Mendelssohn, and himself a great admirer of Beethoven, Bennett developed a piano style that avoided empty display but made considerable demands upon the performer, and maintained most interest in shorter forms. The American MacDowell, like most of his countrymen, received a thoroughly European training that included the encouragement of Liszt and Raff. Though remembered primarily for the *Woodland Sketches* (1896), an amiable series of portraits in the spirit of Schumann, he composed a substantial amount of ambitious music including four sonatas and more than two dozen concert études; the best of this repertory is receiving more frequent hearings today, especially in the USA.

The greater publicity accorded to the French impressionists has served to obscure the unique achievements of Spanish composers at the end of the century. It is easily forgotten that Albéniz's style was already well-formed before Debussy wrote his most important piano works. He enjoyed good relations with both Debussy and Ravel; the influences among the three composers were mutual. Albéniz's major keyboard works, beginning with *La vega* (1897) and culminating in the four books of his *Suite Iberia* (1905–9), were contemporary with important keyboard works of Debussy. Though not as subtle structurally, these pieces are marked by spontaneity and novel figurations, including skilful evocations of both guitar and castanet. Albéniz's countryman Granados excelled in the best tradition of salon music, as in the seven *Valses poeticos*, but his most important publication was the series of *Goyescas* (1911) stimulated by his favourite painter. The best work of Falla and Turina builds upon the achievements of Albéniz and Granados.

Ex.14 Franck: *Prélude, choral et fugue*

Born in the year that Beethoven completed his *Missa solemnis*, the Belgian César Franck did not complete his two most important piano works, the *Prélude, choral et fugue* (1884) and the *Prélude, aria et final* (1887), until Romanticism was about to enter its twilight. In the former especially, he succeeded in tempering a Lisztian technique and cyclic procedures to solemn purpose, often recalling (and almost demanding) an organ pedal board (ex.14). Though greatly influenced by Wagner, Chabrier is often most characteristic in his piano pieces, which contributed in France to the emancipation of dissonance and the interest in modal melodies. Neither Saint-Saëns, Dukas nor D'Indy invested their solo piano music with anything like the interest of their orchestral compositions (and, in Saint-Saëns' case, of his keyboard concertos).

The most important French composer for solo piano in the generation before Debussy was Fauré. Although he cultivated the by now celebrated genres of Chopin (especially the nocturne, impromptu and barcarolle), he brought to each a highly idiosyncratic figuration based upon equal importance of the hands and free polyphony within an arpeggiated background. Unlike much late Romantic keyboard music, Fauré's character-pieces sound less difficult than they are but repay careful study. While Debussy was still writing in a post-Romantic style his contemporary Erik Satie was setting down the three *Gymnopédies* (1888) that, in their sardonic simplicity, helped stake out the composer's iconoclastic position in French musical life. These were succeeded by more than a dozen sets of humorous piano pieces with provocative titles like *Sonatine bureaucratique*; more than his actual music, Satie's acerbic unpretentiousness has exercised considerable influence on 20th-century composers like John Cage.

6. THE GROWTH OF PIANISM, 1900–1940

If the 20th century has so far produced less music for the piano than did the 19th or 18th, the range of its achievement, in terms of widening the expressive potential of the instrument, is notable. With the single exception of Bartók, no composer has contributed to the repertory to anything like the same extent as did Haydn, Mozart, Beethoven or Schubert. Nevertheless, there has been more written for the piano since 1900 than for any other solo instrument, and it is possible to chart the main lines of 20th-century musical thinking from a study of the piano music alone, particularly since a number of composers (including Debussy, Bartók, Schoenberg, Boulez and Stockhausen) have made some of their most important stylistic discoveries through their keyboard works.

Although it may appear that Bartók was the most radical of the early 20th-century composers in attitude to keyboard technique, Debussy, barely a generation his senior, represents an even more fundamental secession from the 19th-century Austro-German pianistic tradition. His imaginative disregard of the essentially percussive qualities of the instrument enabled him to develop a new pianism, dependent on sonority rather than attack, on subtle dynamic shading rather than sustained *cantabile*. His own playing was evidently notable for its range of colour within a *pianissimo* dynamic (aided by the use of both pedals) and this is reflected in a Chopinesque notation that details every nuance of touch, as well as of dynamics and phrasing. Although precise indications of pedalling are rare, Debussy's use of sustained bass notes reveals a new awareness of the possibilities of the sustaining pedal

Ex.15 Debussy: *La terrasse des audiences du clair de lune*

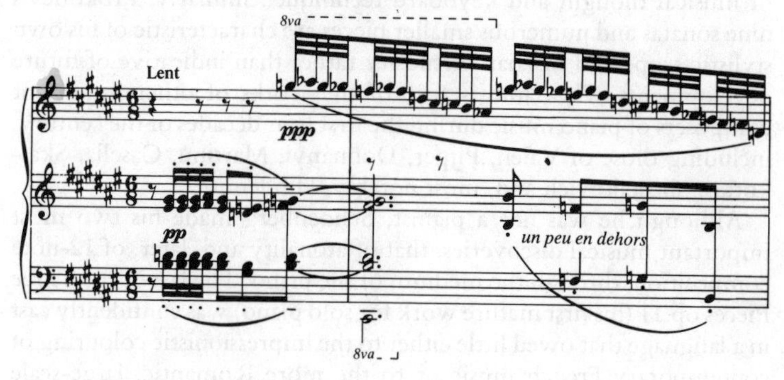

and of the minute differences that can obtain between the total clarity of legato pedalling and the total blurring of undamped strings (see ex.15).

Ravel's more traditional virtuosity, however, marries this new impressionism to a bravura inherited from Liszt, developing a characteristic brilliance of keyboard usage that was, in turn, to have as great an influence on Bartók as did Debussy's more far-reaching experiments in keyboard sonority. As early as 1911, Bartók was stressing the percussive aspect of the instrument through the use of ostinato rhythms; this 'xylophonic' approach was later extended to embrace the more vibraphone-like qualities of a *laissez vibrer* that made expressive use of the suspension and decline of a sound as well as of its initial attack. He was also to continue a Beethovenian investigation of the sharply defined contrasts possible within the instrument's wide dynamic range, and of the contrasts in sound quality suggested by its high, middle and low registers. He continued Debussy's exploration of the resonances obtainable from overlapping harmonies coloured by the sustaining pedal, which later proved equally important in the light of the instrumental techniques proposed by such composers as Messiaen, Boulez and Stockhausen.

Debussy's most important contribution to contemporary pianism resulted from his refusal to acknowledge the essentially mechanical limitations of the instrument, but Ives was to make his contribution through a disregard for the limitations of the ten fingers of the pianist, some of his chords necessitating the assistance of a third hand or of the pianist's arms. If Ives was a prophet ahead of his time, his almost exact contemporary, Rakhmaninov, while making a sizable contribution to piano literature, proved much less significant in relation to the future of musical thought and keyboard technique. Similarly, Prokofiev's nine sonatas and numerous smaller pieces are characteristic of his own stylistic scope and Lisztian virtuosity rather than indicative of future developments. The same is true of the works of other important composers of piano music during the first four decades of the century, including those of Valen, Pijper, Dohnányi, Martinů, Casella, Skalkottas, Shostakovich and, most notably, Hindemith.

Although he was not a pianist, Schoenberg made his two most important musical discoveries, that of atonality and, later, of 12-note composition, through the medium of the piano. The last of the Three Pieces op.11 (his first mature work for solo piano) was confidently cast in a language that owed little either to the impressionistic colouring of contemporary French music or to the more Romantic, large-scale gestures of the late 19th-century Austro-German keyboard composers. The massive stretch of its atonal counterpoints, combined with the extreme contrasts of its fleeting textures and eruptive dynamics (in addition to the exploration of keyboard harmonics in the first piece) remained unique for almost 40 years, until overtaken by still more demanding techniques after World War II. Equally significant in the trend away from Romantic rhetoric, his Six Little Pieces op.19 explore the expressive qualities of the instrument (mostly at the lowest end of the dynamic range) with a restraint more typical of his friend and pupil, Webern, whose single mature work for the piano was such a major landmark. Written during the decade following Schoenberg's inaugural use of the 12-note system in his Suite for piano, op.25, Webern's Variations go even further towards renouncing the grandiose technical obsessions of the recent past – returning to a much earlier conception of instrumental music as an extension of, and almost indistinguishable from, that for voices. The essential simplicity of the piece becomes complex through the continual overlap of wide-ranging contrapuntal lines (and thus of the pianist's hands), demanding a new technical approach to extended part-writing, as well as to the delicate balance between harmonic and rhythmic phrasing (see ex.16).

Ex.16 Webern: Variations, op.27

Webern's piece, with its structural finesse and 'abstracted' *espressivo*, has cast its benevolent shadow on all subsequent composers of piano music.

Stravinsky's pianistic influence extends well beyond the few works he originally wrote for keyboard; not least because he was one of the first composers to establish the piano as an orchestral instrument (Symphony in Three Movements, *Petrushka, The Wedding*). His piano (or piano duet) versions of many of his orchestral works are, in effect, original pianistic conceptions, such was his instinctive feeling for the characteristic spacing of keyboard sonorities.

During the 1920s the American composer Henry Cowell, then regarded merely as an interesting eccentric, began to experiment with hand and arm chord clusters as a means of colouring and outlining his melodic shapes and of creating harmonic 'areas' rather than defined chords. In addition to these keyboard effects he explored the production of sounds directly from the strings themselves, either as pizzicatos; as glissandos on single strings or across the strings (as in *The Banshee*) or in conjunction with silently depressed keys (in order to produce glissando chords, as in *Aeolian Harp*); or as harmonics, produced by the simultaneous 'stopping' of relevant strings.

7. AFTER WORLD WAR II

The possibilities explored by Cowell were woven by John Cage into the aleatory fabric of his most substantial work for piano, *Music of Changes*. Cage also undertook a more radical examination of the piano as a resonating body: the accompaniment to his song *The Wonderful Widow of Eighteen Springs* is rendered entirely on various parts of the frame, which is made to resonate in sympathy with the strings by depressing the sustaining pedal. Moreover, he transformed the basic sound quality of the instrument by a 'prepared' extension of its timbral possibilities: by forcing certain strings to vibrate against wedges made of various materials (metal, wood, rubber etc), Cage opened up a range of keyboard sonorities limited only by the possible damage to the instrument. This, in turn, led to as many variations in the basically harp-like sound of the strings themselves. Robert Sherlaw Johnson has used timpani sticks to stunning effect (in his second sonata), as well as other types of beater; plectra of differing weights and materials have also been used (the eerie sound of a nail-file glissando in Gerhard's *Gemini* is a good example), as have wooden blocks of varying widths (used by David Bedford to produce string clusters in *Piece for Mo*). Bedford has also made weirdly fascinating use of rotating glass milk bottles (at the end of his song *Come in here, child*). The palette of available pianistic colour has continued to be expanded and refined, notably in the chamber works of the American George Crumb, whose (gadgetless) effects are completely viable, often beautiful and never gimmicky.

It seems unlikely, however, that such methods of sound production can become established ingredients of instrumental technique unless manufacturers standardize the shape of the piano's mechanical structure. Because strings freely available on one make of instrument may be hidden or separated by crossbars on another, it is often impossible to carry out the composer's instructions to the letter. In any case, a whole new range of 'string' techniques would need to be developed and practised by the pianist. It is perhaps indicative of the lack of sophistication with regard to these techniques that the two postwar composers to have written the most substantial number of piano

Ex.17 Messaien: *Mode de valeurs et d'intensités*

works, Boulez and Stockhausen, have in general ignored these more peripheral possibilities, as indeed have the majority of other important contributors to the mid-20th-century repertory, including Barber, Sessions, Copland, Feldman, Tippett, Maxwell Davies and Messiaen.

Even the most opulent of Messiaen's recent scores have a muscular background related to the kind of rhythmic counterpoint first developed in his *Mode de valeurs et d'intensités* (1949), in which the basic idea of a rhythmic ostinato was widened into an 'ostinato system' of serial control over the separate elements of duration, dynamics and attack as well as of pitch (see ex.17). This made almost insuperable demands on the performer (as, later, did Boulez's *Structures* for two pianos and Stockhausen's early piano pieces) since such minute degrees of expressive and rhythmic definition, within lines 'broken' by extremes of pitch, are scarcely realizable except by electronic means. The intellectual strictures of Messiaen's early works merged with a freer, unmistakable pianism in his *Cantéyodjayâ* and later in the vast *Catalogue d'oiseaux*, creating a range of keyboard colour as pervasive in its influence on the works of younger composers as was that of Bartók or Stravinsky on the music of an earlier generation.

Boulez's three sonatas and Stockhausen's *Klavierstücke I–XI* (all dating from the late 1940s and 1950s) stand as models of contemporary keyboard writing, as yet unsurpassed in the variety of their neo-virtuosity or in the range of their textural contrasts and expressive sonorities. Musically they display a sharp-edged violence whose stinging contrasts require a comparably 'honed' performing technique – so that these pieces (which had at first seemed largely unplayable) have had the effect of widening the scope and the expectations of virtuoso pianism. They have demanded an increase in pianistic speed and agility in order to encompass complex counterpoints (whether of lines or chords) often involving hand-crossing to an extent that Webern would never have regarded as possible. They also require an ability to define each degree of a dynamic palette that extends from *ppp* to *fff* and beyond, in combination with as many varieties of touch or attack. In the case of Stockhausen, these controls must additionally be linked to an ability to play cluster chords of precisely defined exterior limits, whether these take the form of single attacks, arpeggiated decorations or multiple glissandos (see ex.18, p.142). Moreover, all these new technical requirements involve asymmetrical shapes that need to be mastered as thoroughly as symmetrical scales and arpeggios which again come to the fore in *Klavierstücke XII–XIV* (taken from his opera, *Donnerstag aus Licht*). Unlike the earlier pieces, these make quasi-theatrical use of non-keyboard techniques as well as of certain vocal

Ex.18 Stockhausen: *Klavierstück X*

Clusterglissandi schnell und leicht ohne Rücksicht auf nicht ansprechende Tasten

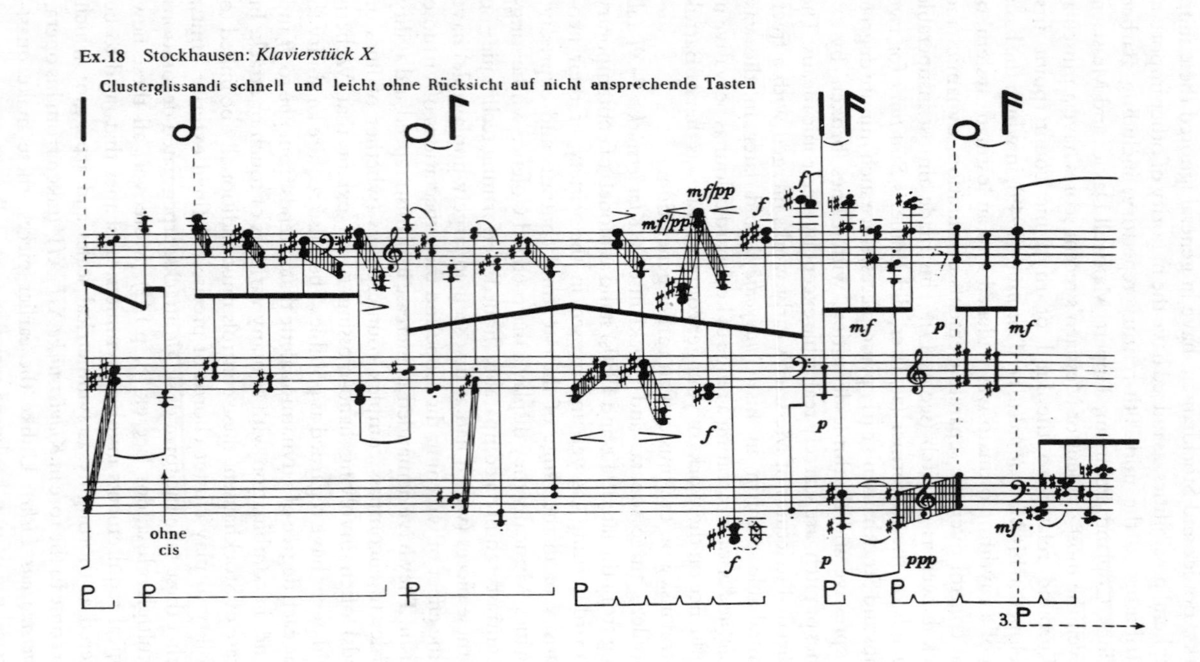

gestures not normally considered part of a pianist's equipment.

In such works, and in pieces by such stylistically diverse composers as Barraqué, Dallapiccola, Berio, Pousseur, Xenakis, Carter and Cage, pedal technique is no longer left to the good taste of the performer but must comply with the specific demands of the score. The use of the sustaining pedal has become as integral to musical expression as dynamics or phrasing: techniques such as half-pedalling, 'after'-pedalling (catching the resonance of a chord after releasing the attack) and 'flutter'-pedalling (effecting the gradual release of an attack) have become commonplace. An increasing number of works (Boulez's Sonata no.3, Stockhausen's *Klavierstücke V–XI* and Berio's *Sequenza IV*, for instance) also require the use of the sostenuto (centre) pedal on concert instruments to free selected strings from the damping mechanism, so allowing them to vibrate in sympathy with any other notes which may be sounded. Berio's *Sequenza IV*, built on the ground bass effect of such sustained notes or chords continually reinforced by the movement of the decorations superimposed on them, provides an ideal study in the use and management of the centre pedal.

APPENDIX ONE

Glossary of Terms

Action. The mechanism of the piano that causes a note to be sounded when a key is depressed. *See* DOWN-STRIKING ACTION, ENGLISH DOUBLE ACTION, ENGLISH SINGLE ACTION, PRELLMECHANIK.

Agraffe. A brass stud, drilled with as many holes as there are unison strings for each note, fixed near the edge of the wrest plank to serve as a bridge for a particular note. It is designed to provide resistance to the displacement of the string caused by the hammer blow and was invented by Erard.

Aliquot scaling. A system of stringing in which unstruck sympathetic strings are added to increase sonority. It is used in the upper register of Blüthner pianos in particular.

Anémocorde [animo-corde, wind-clavier]. A form of SOSTENENTE PIANO in which the vibrations of the strings were sustained by currents of air after receiving an initial impetus from a mechanical bow. It was invented by Schnell and Tschenky in 1789.

Baby grand. The smallest-size GRAND PIANOFORTE.

Back check. *See* CHECK.

Barring. A term for the system of strips or bars of wood glued under the soundboard to enhance its mode of vibration when the strings vibrate (see fig.7*c*); this is crucial to the tone of an instrument.

Bassoon stop. A STOP found on Viennese 18th- and early 19th-century pianos and, rarely, on French. It causes a piece of paper covered in silk to buzz against the bass strings.

Beak. That part of the HAMMER shank in south German and Viennese actions which projects beyond the pivot point and which engages with the escapement mechanism or *Prelleiste* (see fig.4).

Belly rail. The transverse, wooden brace that supports the front edge of the soundboard.

Bentside. The curved side of a piano case.

Bracing. The internal construction of a piano FRAME by which the tension of the strings is supported.

Bridge. A long piece of wood which transmits the vibrations of the strings to the soundboard, to which it is glued. It is shaped to fit the instrument's scaling. The strings pass guide pins, driven into the bridge; these both define the speaking length of the strings and hold them firmly in contact with the surface of the bridge.

Bridle [tape-check]. Part of the action in uprights, patented by Robert Wornum in 1842, which assists the return of the hammer to the check by means of an attached tape, thus facilitating repetition (see p.46, fig.23).

Buff stop [harp stop]. A device found chiefly on square pianos of the 18th and early 19th centuries which muted the strings by pressing a piece of buff leather against them at the nut, so as to damp out high harmonics and give a pizzicato quality to the timbre.

Cabinet piano. A type of upright piano developed by Müller (in Vienna) and Hawkins (London and Philadelphia) in 1800. The case was placed on the floor and the strings passed down vertically behind the action, with the striking point at the top. An essential component of the later cabinet piano was the simple sticker action devised by Southwell in *c*1798. In 1807 Southwell patented a cabinet piano with an escapement action, which continued in production until the 1860s.

Capo tasto [capo d'astro] (It., from *capo*: 'head', *tasto*: 'tie' or 'fret'). The fixed metal bar that exerts a downward pressure on the treble strings of a piano, replacing the wrest-plank bridge (nut); it was invented by Antoine-Jean Bord in 1843. The term is also commonly applied to fretted instruments.

Capstan screw. A regulating screw, with the head cross drilled so that it can easily be turned from the side. An example in modern actions is the screw at the end of the key which transmits the movement of the key when it is depressed, thereby activating the action (see figs.32 and 33).

Check [back check]. A component of grand and upright piano actions. It prevents the hammer from rebounding on to the strings once a note has been struck (see figs.1, 12, 16, 18, 21, 23, 32 and 33).

Chiroplast. A mechanism invented by JOHANN BERNHARD LOGIER for strengthening pianists' fingers and correcting the position of their hands when playing.

Clavecin à maillet (Fr.: 'harpsichord with hammers'). A model for a projected instrument designed in 1716 by Jean Marius to demonstrate the application of a hammer action to a harpsichord. Although there is some confusion with the operating principle of the clavichord tangent at least one of the four such models Marius constructed describes a true piano action.

Clavecin royal (Fr.). A type of square piano invented in 1775 by Johann Gottlob Wagner with unclothed wooden hammers and a number of mechanisms for modifying the basic timbre.

Clavier. (1) (Fr.). A term for the keyboard. (2) A generic term for a keyboard instrument. (3) A term sometimes applied to pianists' practice keyboards that have silent notes ('dumb clavier').

Clutsam Curved Keyboard. Patented by the Australian, George H. Clutsam, in 1907, the Curved Keyboard was shaped so that all notes of the keyboard were equidistant from the player. (Staufer and Haidinger had anticipated this principle in a patent of 1824.) Both Dohnányi and Rudolf Ganz played the instrument, which was manufactured by Ibach in 1908.

Compensation frame. The metal reinforcement of a wooden piano frame, patented in 1820 by James Thom and WILLIAM ALLEN of Stodart, which used brass tubes above brass strings and iron tubes above iron ones so as to maintain greater stability of tuning. The patent marks the turning point in the design of composite frames, in which metal becomes the dominant material of the bracing system. (See also p.41.)

Cottage piano. A generic term for a small upright, also known as a pianino. The name was introduced by ROBERT WORNUM in the early 19th century. (See fig.25.)

Cross-strung. American term for OVERSTRUNG.

Dactylion. A practice device invented by Henri Herz to develop independence of the fingers. It consisted of ten rings, fastened to springs, that hung above the keyboard.

Damper. A part of the mechanism that stops the vibrations of the strings. It is made of felt in modern pianos (previously of leather or cloth) and is mounted on a wooden block. When the key is depressed the damper leaves the string to allow it to vibrate, and as the key is released the damper returns.

Damper pedal. *See* SUSTAINING PEDAL.

Ditanaklasis. Name for the cabinet piano invented by MATTHIAS MÜLLER in 1800.

Doppelflügel (Ger.). A general term for a piano with two keyboards; *see* DOUBLE GRAND PIANO, PIANO À CLAVIERS RENVERSÉS, VIS À VIS.

Double action. *See* ENGLISH DOUBLE ACTION.

Double escapement. A form of ESCAPEMENT.

Double grand piano. A grand piano with two keyboards, one at each end, patented by Pirsson in New York in 1850.

Down-striking action. A horizontal action placed above the strings and frame of a grand or square piano designed to eliminate the two principle disadvantages of a conventional up-striking action: the tendency of the hammer to unseat the string by striking it up and away from the nut; and the structural discontinuity in the piano frame (the gap), provided to admit the hammers. Nannette Streicher's elegant mechanism of 1823 was said to have inspired Henri Pape *c*1827 to develop his own down-striking French grand action, which was to influence the work of a number of European and American makers. Only the invention of the cast iron frame finally made this approach to piano design obsolete.

Duplex scale. A system of stringing introduced by Steinway to improve resonance in the treble. Measured, undamped overlengths of the 'normal' struck unison strings were added so as to vibrate in sympathy with upper partials.

Electrophonic piano. A piano in which electro-magnets, replacing the soundboard, pick up the vibrations of the strings which are then amplified and modified electronically.

Emanuel Moór Pianoforte. A grand piano with two keyboards sounding an octave apart and with an octave coupler; it was introduced by Moór in 1921.

English action. *See* STOSSMECHANIK.

English double action. A development of the English single action that incorporated an intermediate lever and, in Geib's patent, an escapement in the form of a sprung hopper (see fig.11).

English grand action. The action in which the intermediate lever is omitted and a hopper acts directly against the hammer butt. A regulating mechanism ('set-off') disengages the hopper after a note is struck, so that the hammer falls back from the string to be caught and held by the check and so prevented from rebounding. It was developed by Americus Backers (by 1772), with John Broadwood and Robert Stodart, and became standard for English grand pianos (see fig.12).

English single action. The action developed in England in the mid 1700s (chiefly by Zumpe), with the hammer shanks attached to an overhead rail. Two Zumpe square pianos with this action survive from 1766 (see fig.10).

Escapement. Those parts of the piano action (hopper, jack etc) that transmit the motion of the key to the hammer and from which the latter is disengaged just before it strikes the string, allowing it to fall back freely (see fig.5). Cristofori's instrument featured an escapement mechanism (see p.4). In 1821 Erard invented a double escapement, which allowed for the rapid repetition of a note by ensuring that the hammer fell back only a short distance from the string after impact and continued under the control of the key while the key remained depressed. A note could therefore be sounded again without the necessity of completely releasing the key. (See fig.21; *see also* HOPPER.)

Escapement button. *See* SET-OFF BUTTON.

Euphonicon. An upright piano patented in 1841 by John Steward in which the strings stand exposed on a harp-shaped iron frame and the soundboard is replaced by three violin-shaped soundboxes. Its action incorporated Wornum's tape-check principle, in which the tape assists the return of the hammer to the check.

Flügel (Ger.: 'wing'). A term used for the harpsichord and later applied to the GRAND PIANOFORTE.

Flyer [fly lever]. American term for the HOPPER.

Fortepiano. A term sometimes used for the piano of the 18th and early 19th centuries to distinguish it from the more modern instrument. German writers sometimes use the term 'Hammerflügel' for the same purpose.

Frame. The part of the piano that bears the tension of the strings, which are fastened to the wrest plank at one end and to the hitch-pin plate at the other. Iron bracing was introduced to strengthen wooden frames *c*1820 and in 1825 Babcock patented the cast-iron frame, subsequently adopted in all models. *See also* COMPENSATION FRAME.

French grand action. *See* DOWN-STRIKING ACTION.

Gabel-Harmon-Pianoforte. A piano patented by Matthias Müller in 1827 in which one of the unison strings of each note is replaced by a tuning fork.

Gap spacer. Any structural member crossing the gap between the wrest plank and the rest of a piano frame, common before the introduction of full-length iron bracing.

German action. *See* PRELLMECHANIK.

Giraffe piano. A wing-shaped upright with the tail end placed uppermost (see fig.15).

Grand pianoforte. A piano in a horizontal, wing-shaped case, the form of which is directly derived from that of the harpsichord.

Hammer. The part of the piano action that strikes the strings of an individual note (see fig.5). The hammer shanks are usually made of wood; before *c*1830 most hammer heads were covered with leather but since they have mostly been made of densely packed felt (see fig.19). (*See also* BEAK.)

Hammerflügel (Ger.). *See* FORTEPIANO.

Hammerklavier (Ger.). A term for the piano used in Germany in the late 18th and early 19th centuries: it was designated by Beethoven for his sonatas opp.101 and 106.

Hand stop. *See* STOP.

Harp stop. *See* BUFF STOP.

Hitch-pin. The metal pin which secures the strings at the end opposite to the wrest pin. On the grand piano the hitch-pins are at the end furthest from the player; their position on other pianos varies according to structural factors.

Hitch-pin plate [hitch-plate]. The plate into which the hitch-pins are driven.

Hopper [grasshopper, jack flyer, fly lever, flyer]. A pivoted or hinged jack that permits a hammer to 'escape' and fall back from the string while the key remains depressed (see fig.12). *See also* ESCAPEMENT.

Jack [pilot]. A small rigid upright with a leather button on top, fixed to the key (see fig.1); it transmits the movement of the key to the hammer or an intermediate lever. Zumpe's double action instruments feature two jacks (or pilots) – one attached to the key which acts on an intermediate lever, and the other attached to the intermediate lever which acts on the hammer. It may form part of the ESCAPEMENT mechanism (see fig.32).

Janissary stop [Turkish music]. A pedal that operated a drumstick, which struck the underside of the soundboard, tuned bells and a cymbal to add 'Turkish' effects fashionable in late 18th- and early 19th-century Vienna.

Kapsel (Ger.). A wooden or metal block or fork within which the hammer is pivoted in the *Prellmechanik* (see pp.13 and 18; figs.5 and 18).

Key. A balanced lever which when depressed operates the action.

Keyboard. A set of levers that actuates the mechanism or action of instruments such as the piano.

Key frame. The wooden frame fitted with guide and balance pins to which the keys are attached.

Knee lever. A device operated by the knee that was the predecessor of the PEDAL on early continental pianos. It could be used to lift the dampers, to engage a muting device or to shift the action.

Liner. A wooden rail glued round the inside of the piano case or rim to which the soundboard is glued.

Loud pedal. *See* SUSTAINING PEDAL.

Lyraflügel (Ger.). LYRE PIANO.

Lyre. The wooden frame, often decorated, which supports the pedals of a grand piano.

Lyre piano. A variant of the pyramid piano, current from *c*1830; the case was in the form of a lyre.

Minipiano. A very small upright with a 'drop' action (with the mechanism placed below the level of the keyboard), first produced in England in 1934.

Moderator [muffler pedal]. A knee lever or pedal that introduces a strip of cloth between the hammers and the strings to produce a muted effect. It was important on 18th- and early 19th-century Viennese pianos and provided a special sonority; it is not found on English instruments of the period and rarely on French. On modern uprights, as a third pedal, it is employed to reduce volume while practising.

Nachttisch (Ger.: 'night table'). A small version of the square piano.

Nut [wrest-plank bridge]. The bridge normally positioned nearest the wrest pins and opposite the soundboard bridge.

Orphica. A portable piano, designed for outdoor use, patented in 1795 by K. L. Röllig. It could be played either resting on the player's lap or strapped around his neck.

Overdamper. A term applied to upright pianos in which the dampers are above the hammers.

Overspun string. A metal string that has a thin ductile wire wound around it, so as to increase its mass while avoiding the loss of flexibility found in a plain wire of the same mass. The bass strings of a piano are overspun with copper wire.

Overstrung. A term applied to a piano in which the strings are arranged in two nearly parallel planes, with the bass strings passing diagonally over those of the middle range (see figs.27, 28 and 29*b*).

Pedal. A foot-operated device that either activates a particular component of the piano action or modifies its mode of operation so as to alter the tone-colour or volume and produce expressive effects. 18th- and early 19th-century pianos often had a variety of pedals or hand stops, but on modern instruments there are usually only two or three. *See* SOSTENUTO PEDAL, SUSTAINING PEDAL, UNA CORDA.

Pedal-board. The keyboard played by the feet, connected either to the strings of the manual keyboard or to a separate soundboard and strings attached to the underside of the instrument.

Pedalier. (1) A pedal keyboard attached to a piano and capable of activating its hammers. (2) An independent PEDAL PIANOFORTE made by Pleyel, Wolff & Cie, to be placed underneath an ordinary grand piano.

Pedal pianoforte. A piano with a pedal-board like that of an organ. Three types of 18th-century pedal pianos are known. An instrument by Johann Schmidt (in the Metropolitan Museum of Art, New York) has pedal-operated hammers which strike the same strings as those struck by the manual hammers; another has a separate soundboard and strings attached to the underside of the instrument. The third type has an independent pedal piano placed under a grand piano; no 18th-century example is known, but such instruments became the standard type in the 19th century (an instrument, *c*1815, by Brodmann is in the Kunsthistorisches Museum, Vienna).

Pianette. A very low pianino, or upright piano, introduced by Bord in 1857.

Pianino (It.: 'small piano'). A small upright, originally designed by Pape and introduced by the Pleyel firm; it is the continental equivalent of the COTTAGE PIANO.

Piano (Ger.). UPRIGHT PIANOFORTE.

Piano à claviers renversés [Piano Mangeot] (Fr.). A double grand piano, originally designed by Jozef Wieniawski and patented by E.J. Mangeot in 1876, with two keyboards, one above the other; the ascending scale of the upper one ran from right to left.

Piano a coda (It.). GRAND PIANOFORTE.

Piano à prolongement (Fr.). A piano in which sustained sounds could be produced by the simultaneous vibration of free reeds set in motion when the strings were struck in the usual way by hammers. It was made in Paris by Alexandre in the 19th century. (*See also* SOSTENENTE PIANO.)

Piano à queue (Fr.). GRAND PIANOFORTE.

Piano attachment. A small electronic keyboard that imitates orchestral instruments, normally used as a melody instrument accompanied on the piano with the left hand. It is clamped to, or placed in front of, a piano keyboard with notes of the same pitch aligned. It usually has a range of three octaves and can be set within a total range of five or six octaves. It has between 12 and 22 stop-tabs to control timbre, attack and vibrato. It was invented in the late 1930s and popular until the mid-1950s.

Piano carré (Fr.). SQUARE PIANOFORTE.

Piano droit (Fr.). UPRIGHT PIANOFORTE.

Piano éolien (Fr.: 'aeolian piano'). A keyboard instrument in which the vibrations of the strings, which are activated by hammers, are sustained by jets of compressed air.

Pianoforte. Name for the piano that derives from the instrument's capability of being able to sound *piano e forte* ('soft and loud'), according to touch.

Pianola. An automatic piano-playing device (*see* PLAYER PIANO) invented in 1895 by Edwin Scott Votey and made by the Aeolian Corporation. The trademark 'Pianola' is frequently misapplied to instruments of other makes.

Pianoline. A PIANO ATTACHMENT, first made in 1950 by Lipp. It was derived from G. Jenny's Ondioline.

Piano Mangeot. *See* PIANO À CLAVIERS RENVERSÉS.

Piano player. Part of the apparatus used in the original form of the PLAYER PIANO.

Piano scandé (Fr.). A sostenente piano in which sustained sounds could be produced by the simultaneous vibration of free reeds set in motion when the strings were struck in the usual way by hammers. It was invented in 1853 by Lentz and Houdart.

Piano trémolophone (Fr.: 'tremolo piano'). A sostenente grand piano with two keyboards, of which one was exclusively for tremolando notes produced by quick repeated movements of the hammers. It was made by Philippe de Girard, who patented it in 1842. Similar devices were experimented with as early as 1800 by Hawkins and others.

Pilot. *See* JACK.

Pinblock. American term for the WREST PLANK.

Player piano. A piano which automatically plays music recorded, usually, by means of perforations in a paper roll. In the 1890s, the mechanism (piano player) was in a separate cabinet, pushed in front of an ordinary piano. The roll passed over a metal tracker bar with a slot for each note. When a perforation uncovered a slot, suction, generated by pedals, operated a pneumatic valve and lever, forcing down a wooden 'finger' that projected over the piano keyboard. Tempo, dynamics and the sustaining pedal of the piano were controlled by levers in the front of the cabinet. About 1900 the piano player was built into the piano, with control knobs along the front of the keyboard and pumping pedals underneath. 'Expression' pianos, including the Aeolian 'Themodist' (*c*1906), equipped to interpret expressive effects incorporated directly into the music roll, were also produced. The most sophisticated form of the instrument

– the reproducing player piano – re-created the nuances in performance of artists such as Paderewski and Rakhmaninov; mechanisms were usually powered by electricity. Well-known models included the 'Welte-Mignon' (devised by Edwin Welte, 1904), the 'DEA' (Hupfeld, 1905), the 'Duo-Art' (Aeolian Co., 1913), and the 'Ampico' (American Piano Co., 1916). The Welte-Mignon was available in pianos of 115 different makes, including Steinway and Gaveau. Production reached its peak by 1923 with nearly half a million player pianos manufactured in two years, but their popularity declined largely as a result of the Depression and the increased use of the radio and gramophone.

Prelleiste. A fixed rail, in the primitive *Prellmechanik* without escapement, with which the hammer beaks engage when a key is depressed (see fig.4).

Prellmechanik. The German and Viennese type of piano action, in which the hammer heads face towards the player and the hammer shanks are pivoted directly on (or held in a *Kapsel* attached to) the key. It was developed with escapement by such makers as J.A. Stein in Germany and by Walter, Nannette Streicher and M. A. Stein in Vienna and became known as the Viennese action. (See pp.12–15 and figs.4, 5, 8, 16 and 18.)

Prepared piano. A piano in which the pitches, timbres and dynamic responses of individual notes have been altered by means of objects such as bolts, screws, mutes or rubber erasers inserted at particular points between the strings. The technique was developed by the American composer John Cage from *c*1940. (See fig.36.)

Pyramid piano. An early type of upright piano in which the strings run at a slight diagonal, permitting them to be enclosed in a symmetrically tapering case.

Querflügel [Querpianoforte] (Ger.). A small horizontal piano, usually wing-shaped with the keyboard placed at an acute angle to the spine. The wrest pins are positioned behind the keyboard as in a grand but with the strings running diagonally to the right. The type is found from at least the 1780s.

Regulating button [regulating screw]. An adjustable screw used to alter various required tolerances throughout the piano mechanism.

Scaling. The relationship between the sounding length of the strings and their pitch. Modern pianos have scaling that is gradually shortened throughout the range from treble to bass.

Set-off button [escapement button, set-off screw]. A regulating button that controls the moment (the 'set-off') at which a jack or hopper is made to disengage from the hammer when a key is depressed (see figs.21, 23, 32 and 33).

Short octave. A term to denote the lowering of pitch of some bass notes to extend the compass downwards. This was to meet the requirements of a specific piece of music and was not practised after the early 19th century.

Single action. *See* ENGLISH SINGLE ACTION.

Soft pedal. *See* UNA CORDA.

Sostenente piano. A term used in connection with attempts to produce a strung keyboard instrument that could sustain sound. In the 18th and 19th centuries a number of devices were invented, often taking the form of attachments to the action. They included continuous bows, bellows through which jets of air were driven against the strings, vibrating rods, free reeds and tremolo mechanisms. The modern sostenente piano is the 'electronic' piano.

Sostenuto pedal. A middle pedal found on some grand pianos that enables the player to sustain the sound of one or more notes held down at the moment the pedal is depressed, without removing the dampers from all the strings of the instrument (*see* SUSTAINING PEDAL.).

Soundboard. The thin sheet of wood to which the bridge is glued. The bridge transmits the vibrations of the strings to the soundboard which, with its vastly greater surface area, sets the air into vibration with increased efficiency, thereby producing a louder sound. The energy of the string is dissipated more rapidly and is converted to a sound of higher intensity that lasts for a shorter time. The thickness and shape of the soundboard (and therefore its vibrational characteristics and resonance) are important in determining the tone quality of the instrument.

Spine. The long side of a grand piano case, opposite to the bentside.

Spinet. A term used in the USA during the 1930s to describe a miniature upright piano.

Square pianoforte. An early piano that retained the rectangular shape of the clavichord. (See figs.8, 9, 14, 24 and 27.)

Sticker. A rigid rod, usually wooden, which exerts a pushing action against either the hammer or damper in a piano mechanism (see figs.10 and 22; *see also* STICKER ACTION).

Sticker action. The action developed initially by Southwell for an upright version of a square piano and later applied to the English upright cabinet piano. A sticker, sometimes of great length, is used to communicate the motion of the hammer to the jack at the top of the case.

Stop. A hand control, knee lever or pedal that mechanically modifies the piano's timbre.

Stossmechanik [English action] (Ger.). A piano action in which the hammer, pointing towards the back of the instrument, is hinged to a fixed rail and is thrust upwards by a jack or escapement mechanism attached to the key lever. The Stossmechanik in which the hammers point towards the player (as in the *Prellmechanik*) is normally classified as Anglo-German but on the continent was often called the English action.

Strahlenklavier (Ger.: 'beamlike piano'). A redesigned piano, made by Bechstein *c*1870, which divided the keyboard into two arcs to allow for the range of each arm. The design was faulty, since it resulted in the player constantly having his elbows pressed into the body.

Support. *See* WIPPEN.

Sustaining pedal [damper pedal, loud pedal]. The right-hand pedal of the modern piano which when depressed raises the dampers from all the strings, allowing them to vibrate freely until the pedal is released. (*See also* SOSTENUTO PEDAL.)

Sympathetic strings. Unstruck strings that are added so as to increase sonority. *See* ALIQUOT SCALING; DUPLEX SCALE.

Tangent piano (Ger. *Tangentenflügel*). A keyboard instrument, said to have been invented by Späth *c*1751, in which the strings are struck by freely moving slips of wood resembling harpsichord jacks. Späth's design was anticipated by Schröter in 1739 and possibly by Marius in 1716.

Tape-check. *See* BRIDLE.

Telio-chordon. A microtonal grand piano invented by Charles Clagget in 1788. The octave was divided into 39 intervals and each key, with the help of pedals, operated three notes.

Touche (Fr.). KEY.

Transposing keyboard. A keyboard that can be shifted laterally, usually by means of a lever, enabling a performer to play music in a different key from that in which it is written. The keys, when shifted, strike strings at a chosen interval above or below the norm.

Tuning-pin. *See* WREST PIN.

Turkish music. *See* JANISSARY STOP.

Una corda [soft pedal] (It.: 'one string'). The left-hand pedal on the modern piano. In 18th- and early 19th-century English grands and early English uprights it caused the action to shift so that (if the pedal is fully depressed) only one string per note was struck. No such pedals are found on south German and early Viennese pianos. Later Viennese pianos foreshadowed modern grands, in which only two of the three strings are struck in the treble, and one of the two in the bass. On an upright the 'soft pedal' moves the hammers closer to the strings.

Underdamper. A term applied to upright pianos in which the dampers are below the hammers.

Upright grand pianoforte. A grand piano enclosed in a rectangular case and placed upright on a stand; it was introduced by Robert Stodart in 1795.

Upright pianoforte. A piano constructed in a vertical case. The prototype of the modern piano in which the strings pass behind the keyboard to the bottom of the case, resting on the floor was created by Hawkins in 1800; Müller's 'Ditanaklasis' (1800) also had vertical stringing. (See pp.14–15; *see also* CABINET PIANO; COTTAGE PIANO; GIRAFFE PIANO; LYRE PIANO; PYRAMID PIANO.)

Viennese action. *See* PRELLMECHANIK.

Vis à vis (Fr.). A term applied to a keyboard instrument with two keyboards, one at each end.

Voicing. The means by which a particular quality of timbre, loudness etc is achieved. It involves altering the hardness of the sub-surface felt of the hammers by pricking them with needles once they have been shaped.

Wippen [support]. A pivoted lever which, as part of the action, raises the jack when a key is depressed (see figs.32 and 33).

Wrest pin [tuning-pin]. The steel pin, driven into the wrest plank, around one end of which a string is wound. The pin is turned with a tuning key to alter the tension of the string, and thereby its pitch.

Wrest plank [pinblock]. The piece of wood into which the wrest pins are driven. In early instruments it was usually made of solid beech, walnut, maple or oak but now it is of cross-laminated maple and is supported by the cast-iron frame.

Zuggetriebe [Zugmechanik] (Ger.). A term coined by W. Pfeiffer to describe a German action in which the motive force is transmitted to the hammer beak by a mobile escapement lever.

APPENDIX TWO

Index of Makers

Aeolian Co. American firm of player piano makers. It was founded in 1878 by William B(urton) Tremaine (1840–1907), originally a piano builder, as the Aeolian Organ & Music Co. to manufacture automatic organs and perforated music rolls. His son Henry B(arnes) (1866–1932) expanded the company into one of the largest manufacturers of player pianos: during the first three decades of the 20th century it sold millions of instruments. It introduced the sophisticated Duo-Art Reproducing Piano in 1913 and manufactured several Pianola models. In 1903 the Aeolian Weber Piano & Pianola Co. was formed of which Aeolian Co. was a major part. The company eventually controlled many important firms, including the Choralian Co. (Germany), the Orchestrelle Co. (England), Mason & Hamlin, Steck and the Weber Piano Co. In 1932 it merged with the American Piano Corporation to form the Aeolian American Corporation.

Aeolian Corporation. American firm. It was formed in 1932 as the Aeolian American Corporation as the result of the merging of the Aeolian Co. and the American Piano Corporation. Its assets were bought in 1959 by Winter & Co., which changed its name to Aeolian Corporation in 1964. The firm acquired the assets of many formerly independent piano companies and manufactured instruments under the following trade names: Mason & Risch (Toronto); Mason & Hamlin; Chickering; Wm. Knabe; Cable; Winter; Hardman, Peck; Kranich & Bach; J. & C. Fischer; George Steck; Vose & Sons; Henry F. Miller; Ivers & Pond; Melodigrand; Duo-Art; Musette; and Pianola Player Pianos. In 1980 the name was changed to Aeolian Piano, Inc.; it was purchased by Peter Perez (a former president of Steinway) in 1983, but went bankrupt in 1985. Wurlitzer then acquired certain assets, including the trade names Mason & Risch, Chickering, Cable, Winter, J. & C. Fischer and Henry F. Miller.

Albrecht, Charles (1759/60–1848). American maker of German birth. Active in Philadelphia, *c*1790–1824, he made some of the earliest surviving American square pianos, with a range of five to five and a half octaves and handsome cabinet work.

Albrecht, Christian Frederick Ludwig (1788–1843). American maker of German birth. Active in Philadelphia from *c*1822, he made upright and square instruments of exemplary quality, with a six-octave compass.

Albrecht & Co. American firm. It was formed in Philadelphia as the result of partnerships set up between 1864 and 1876 by Charles Albright (Albrecht from 1864); from 1887 to 1916 it was owned by Blasius & Sons, the name continuing until 1920.

Allen, William (*fl* London, *c*1820–31). English maker. He worked in William Stodart's workshop in London and with James Thom invented the 'compensation' frame for grand pianos (patented 1820), which used metals compatible with the strings to prevent fluctuations in tuning.

American Piano Co. American firm. It was incorporated in 1908 and consolidated firms such as Chickering and Knabe with the Foster–Armstrong Co. of East Rochester (NY). It marketed pianos ranging from concert to mass-produced commercial instruments and in 1909 established a player-piano department. The firm's Ampico became one of the most sophisticated reproducing player mechanisms. In 1922 Mason & Hamlin became part of the firm but was sold to the Aeolian Co. in 1930. The American Piano Co., from 1930 the American Piano Corporation, merged with the Aeolian Co. to form the Aeolian American Corporation in 1932.

Babcock, Alpheus (1785–1842). American maker. He invented the single-cast metal frame with hitch-pin plate (patented 1825), from which the modern frame was developed. His square pianos were of superb quality with a range of five and a half or six octaves. Babcock worked for or was a partner in several firms in Boston and Philadelphia, including Babcock, Appleton & Babcock (1811–14), John and G. D. Mackay (1822–9), William Swift (1832–7) and Chickering (1837–42).

Backers, Americus (*fl* 1763–78). Dutch maker. Active in London, he is credited with inventing the English grand action by 1772 (see p.25). He also made harpsichords.

Baldwin. American firm of instrument makers, especially of pianos and organs. D. H. Baldwin began manufacturing in Cincinnati in 1862; branches opened later in Louisville and Indianapolis. From the 1860s to the late 1880s the firm was one of the largest dealers in the Midwest and by 1891 the Baldwin Piano Co. was manufacturing low-priced upright pianos. From 1903 the Wulsin family controlled the firm and by the 1970s also marketed a range of electronic instruments, including pianos. The corporation took over Bechstein in 1963. Baldwins have a high reputation in the USA, but are less well known in Europe.

Ball, James (*fl* 1780–*c*1832). English maker. Also a music publisher; he worked in London. His five-octave square pianos had the English single action, and he also made cabinet pianos and grands, patenting various improvements which were not adopted.

Bechstein. German firm. It was founded in Berlin in 1853 by Wilhelm Carl Bechstein. By the late 1860s Bechstein grands were widely admired and the firm's annual output rose to 5000 instruments by 1914. After 1870 all were iron-framed and overstrung. The Bechstein of *c*1900–14, with its 'velvety' bass and expressive treble tone (less brilliant than that of Steinway) became the favoured piano of many leading artists; the model 8 upright (pre-World War I) and model 9 (1930s) are among the finest ever made (see fig.30). An attempt to market the electronically amplified 'neo-Bechstein' from 1933 failed. The factory was destroyed in World War II but the firm recovered fully and has maintained its high reputation. It was taken over by Baldwin in 1963.

Beck, Frederick (*fl* 1756–98). German maker. Active in London, he made square pianos of very variable quality; the finest date from his early years. His surviving instruments have a range of five octaves or slightly less, a single action, two or three hand stops and very rounded hammer heads covered with soft leather.

Becker. Russian firm. From 1841, when it was founded in St Petersburg, to 1891 over 11,000 instruments were produced. Becker adopted the American system of cross-stringing and by 1865 Erard's repetition action. The grands were used by many leading virtuosos, including Anton Rubinstein. Production ceased in 1914.

Belt, Philip R(alph) (*b*1927). American fortepiano maker. Since 1966 he has made several replicas (some in kit form) of instruments by such makers as Stein and Walter; they have a clear sound and fluid touch.

Beregszászy, Lajos (1817–91). Hungarian maker. From 1846 until 1879 he produced many quality pianos in Budapest which were sold throughout Europe. His patent for a curved sounding board was bought by Bösendorfer.

Biese, Wilhelm (1822–1902). German maker. He established a firm in Berlin, *c*1850, which specialized mainly in uprights, producing 15,000 instruments by 1893. It continued manufacture into the 1970s.

Blanchet. French family, active in Paris. They are renowned for their harpsichords, though from *c*1830 specialized in high-quality small uprights. PASCAL TASKIN worked with the firm and on the death of François Etienne Blanchet in 1766 took over the workshop. François' son Armand François Nicolas (1763–1818) worked with Taskin during the firm's transition to piano making. Jean Roller became a partner of Armand's son Nicolas, manager from 1818. The firm built its first upright in 1827; its improved model (1830) was widely imitated.

Blüthner. German firm. It was founded in Leipzig in 1853; initially building only grands, it made uprights from 1864. In 1873 the aliquot scaling of grands was patented, which added a sympathetic string to each trichord in the treble to enrich sonority. The factory was destroyed in World War II, but the firm has since regained its former eminence and it opened a new factory in 1974. Blüthner instruments are distinguished by a romantic tone and full treble.

Boisselot. French firm. Founded in Marseilles in 1828, it existed until the end of the 19th century. Its innovations included a patent (1843) for a piano with two sympathetic octave strings (predating Blüthner's 'aliquot' scaling) and a piano with a *pédale tonale* (exhibited 1844), similar to the sostenuto pedal introduced by Albert Steinway in 1874 for square pianos.

Bord, Antoine-Jean Denis (1814–88). French maker. In 1843 he invented the *capo tasto*. The firm made small, cheap and durable uprights, but these were later supplanted by German overstrung iron-framed instruments. In 1934 the firm was taken over by Pleyel.

Bösendorfer. Austrian firm. It was founded in Vienna in 1828 by Ignaz Bösendorfer (1796–1859), a former apprentice to Brodmann. Liszt's approbation of Bösendorfer's instruments brought the firm fame. Between 1828 and 1975 it marketed 33 different models, initially making both English- and Viennese-action instruments, but after 1910 almost exclusively the former. The concert grands, especially the 'Imperial' with a range of eight octaves, are noted for rich overtones (ideal for chamber music). A limited number of fine uprights were made. In 1909 the firm passed to the Hulterstrasser family and in 1966 it was taken over by Kimball.

Brinsmead. English firm. Founded in London in 1835, it became a limited company in 1900, with an annual output of *c*2000 pianos of above average quality. In 1967 Brinsmead was bought by Kemble.

Broadwood. English firm. It was founded in London by the cabinet-maker John Broadwood (1732–1812), who became the partner of the harpsichord maker Burkat Shudi in 1770. Broadwood gained control of the firm after Shudi's death. His first square pianos were based on Zumpe's models, but in 1783 he patented a wholly redesigned instrument with wrest plank and pins at the back (see p.10). Broadwood had earlier worked with Backers and Stodart on the English grand action. He then began to design grands (at first after Backers) and during the next few decades various innovations to increase the power, dynamic range, compass and durability of the instruments were adopted, including the addition of iron bracing to grands (*c*1820). This was developed by H. F. Broadwood into the 'iron grand'. By 1851 annual output had reached 2500 and the firm had attained considerable prestige but thereafter a failure to adopt new technology led to its decline, although it is still active.

Brodmann, Joseph (1771–1848). Austrian maker. Well established in Vienna by 1796, he built a 'Querpianoforte', a piano with the keyboard placed at an acute angle to the end of the instrument; he also built independent pedal keyboards. His workshop was bought by Ignaz Bösendorfer on his retirement in 1828.

Challen. English firm. Founded in London in 1804, Challen has specialized in small grands, with a reputation for good quality pianos at a reasonable price. The firm was bought by Barratt & Robinson in 1971.

Chappell. English firm. Initially active in London as publishers, it made pianos from *c*1840 and by 1893 had produced over 30,000, mostly quality uprights. It subsequently bought Allison and Collard & Collard. Chappell pianos are now made by Kemble.

Chickering. American firm. It was founded as Stewart & Chickering in Boston (1823) when Jonas Chickering (1797–1853) went into partnership with JAMES STEWART. When Stewart left for England Chickering worked alone and then from 1830 to 1841 formed a business relationship with John Mackay. The firm at first made square pianos but was also producing uprights by 1830 and grands by 1840. In 1843 Chickering patented a one-piece cast-iron frame, his most significant innovation and one that became the basis of piano manufacture. After Jonas's death the firm was continued by his sons and their descendants, winning many awards in the USA and Europe. It was bought by the American Piano Co. in 1908 and became part of the Aeolian American Corporation in 1932. In 1985 it was acquired by Wurlitzer.

Clagget, Charles (1740–*c*1795). Irish violinist and inventor. He devised the 'telio-chordon', a piano in which each octave was divided into 39 intervals, each key giving three pitches for any given note.

Clementi. English firm. It was founded in London by the composer Muzio Clementi, who in 1798 went into partnership with Longman, Collard and others. Clementi did much in Europe to promote sales of their pianos, which initially kept the basic designs of Longman & Broderip. Clementi introduced the six-octave square piano *c*1810 and incorporated modifications made by JAMES STEWART in 1827. After Clementi's retirement the firm continued as COLLARD & COLLARD. (See also figs.24 and 25.)

Collard & Collard. English firm. It descended from Longman and his associates and continued the London business of CLEMENTI. By the mid-19th century it was second only to Broadwood in output, but was slow in adopting new technology. Collard absorbed the Kirckman business in 1896. In 1929 it was taken over by Chappell, the name continuing until 1971. Collard instruments have a full and pleasant tone.

Cramer, J. B. English firm. It was founded in 1824 in London as Cramer, Addison & Beale when the pianist and composer Johann Baptist Cramer joined the partnership of Robert Addison and Thomas Frederick Beale. Until Cramer's death in 1858 it was chiefly concerned with music publishing; thereafter pianos were manufactured with George Wood as Beale's partner. In the 20th century publishing again became the main activity; in 1964 the firm passed to Kemble.

Crehore, Benjamin (1765–1831). American maker of string and keyboard instruments. Working in Milton (Mass.), he was especially renowned for his pianos (from 1797) and founded the New England industry, training Alpheus Babcock and others. His five surviving square pianos have a range of five to five and a half octaves, Zumpe action, and a long soundboard extending across the key frame.

Cristofori, Bartolomeo (1655–1731). Italian maker. He worked in Florence from 1690 and invented the piano in about 1700. His instruments had leather-covered hammers and an inverted wrest plank with the tuning-pins running through it, the strings being attached to the underside (see pp.4–8 and figs.1 and 2). He also made harpsichords.

Dietz, Johann Christian (*c*1804–88). German maker. Like his father (also Johann Christian), he invented various instruments. He designed a grand piano with freely vibrating sides to the soundboard, and another with a fourth, undamped string for each note which provided greater resonance.

Dodds & Claus. American firm. Founded by the Englishman Thomas Dodds and the German Christian Claus, it established the New York piano industry, 1791–3. Square pianos were made on the Broadwood model, but with inferior workmanship.

Dolge, Alfred (1848–1922). German manufacturer. He emigrated to the USA in 1866 and became noted for his high-quality piano felts and soundboards, and as an importer of piano supplies. He maintained a business in New York.

Ekström. Swedish firm. Founded in 1836 in Malmö, it established a reputation for quality instruments. In 1967 it was taken over by A. B. Nordiska of Vetlanda.

Erard. French firm. It was founded in Paris by Sébastien Erard (1752–1831), who used a Zumpe square piano as the model for his first instrument. He subsequently made five-octave pianos and developed an instrument with three strings per note and a double pilot. Large pianos with a range of up to seven octaves were made although most had six and a half octaves. Erard patented both the agraffe and the highly important double escapement action of 1821 (see also pp.29 and 41–3 and fig.21). In 1792 Sébastien opened a London workshop, later devoted to harps, while the Paris firm concentrated on uprights. After Sébastien's death his nephew Pierre took over both divisions. From 1903 to 1959 the firm traded as Blondel et Cie (Maison Erard), Guichard et Cie (Maison Erard) and Erard et Cie. It amalgamated with Gaveau in 1960 and Erard-Gaveau merged with Pleyel in 1961; the firm was taken over by Schimmel in 1971. Winners of many awards, Erard pianos were noted for their ringing tone; Paderewski favoured Erard instruments.

Feurich. German firm. It was founded in 1851 by Julius Feurich (1821–1900), and originally concentrated on making uprights, although latterly it has made concert grands, which are highly regarded. The present factory is in Langlau, Bavaria.

Fischer, J. & C. American firm. It was founded in New York in 1840 by the brothers John N. and Charles S. Fischer, who were of Italian origin. The firm made both uprights and grands and for a time was associated wth William Nunns. It was incorporated in 1907 and later became part of the American Piano Co.; Wurlitzer acquired the firm in 1985.

Förster. German firm. It was founded by F. A. Förster in Löbau in 1859, by the 1880s producing *c*500 uprights annually. In 1924–5 the company devised the quarter-tone piano and in 1933 patented an electrophonic piano. The firm has maintained a steady production of medium-quality instruments.

Foster–Armstrong Co. American firm. It was founded in Rochester (NY) by George G. Foster and W. B. Armstrong in 1894 and in 1899 bought the Marshall & Wendell Piano Co. of Albany. After the construction of a new plant in 1906 it acquired other firms and in 1908 became part of the AMERICAN PIANO CO.

Ganer, Christopher (*fl* 1774–1809). German maker. Active in London, he made square pianos with a single action and overdampers, having a range of around five octaves. His instruments vary in decorative quality.

Gaveau. French firm. It was founded in Paris in 1847 and also made harpsichords. The firm's reputation and large output was due to a sound commercial sense. In 1960 Gaveau joined with Erard; Erard-Gaveau merged with Pleyel in 1961 and in 1971 was taken over by Schimmel.

Geib. Family of piano makers of German origin. John Geib (1744–1818) ran a successful factory in London, making *c*5400 instruments; he claimed to be the first to make 'organized pianos'. In 1786 he patented a double action, which used the hopper for the first time (see p.25). In 1797 he settled in New York as an organ builder. In 1818 his sons J. A. and W. Geib were established as a firm making square pianos.

Graf, Conrad (1782–1851). German-born maker. He established his own workshop in Vienna in 1811 and became piano maker to the royal court in 1824; he was less an innovator than a fine craftsman. A typical Graf has an essentially wooden construction, a range of six and a half octaves and three to five pedals, with a slightly heavier version of the Viennese action. His pianos were used by Beethoven, Liszt, Thalberg and Schumann (see fig.17).

Grotrian-Steinweg. German firm. It derived from a branch of the Steinweg firm that remained in Germany when Theodor Steinweg moved to New York (*see* STEINWAY). Friedrich Grotrian (1803–60) became a partner in 1858; his son Wilhelm took control in 1886, since when the firm has remained in the family. Widely admired by such pianists as Clara Schumann and Gieseking, its pianos are sonorous and tonally refined.

Haines Bros. American firm. It was founded in New York in 1851 by the brothers Napoleon and Francis Haines, who had worked since 1839 with the New York Piano Manufacturing Co. They were among the first in the USA to produce modern overstrung instruments. The company later became part of the Aeolian Corporation, which went out of business in 1985.

Hals. Norwegian firm. It was established in Christiania in 1847. By 1869 uprights were made with a seven-octave compass and iron bars and metal plates as bracing; grands were also manufactured. The company was taken over by Grøndahl in 1925.

Hardman, Peck. American firm. It was founded in New York in 1842 by Hugh Hardman, who had emigrated from Liverpool in 1815. The company took out patents for a key frame support and for a 'harp stop', and made mini pianos under licence in the USA. It later became part of the Aeolian Corporation.

Hawkins, John Isaac (*fl* 1799–1845). English maker and inventor. Working in London and the USA, he invented an upright piano (contemporaneously with Müller in Vienna), which he patented in 1800 (see p.43); its action is in essence retained today. His instruments, although cheap, were never popular, possibly because of inferior tone.

Haxby, Thomas (1729–96). English instrument maker. He was active in York; his surviving square pianos, based on the earliest English models with up to five octaves, show a craftsman of superior taste and technical skill.

Heilman, Matthäus (1744–1799 or 1817). German maker. Trained possibly under J. A. Stein, he settled in Mainz. As a technician working on Stein's instruments he claimed that his own were more reliable.

Herz, Henri (1803–88). German maker, also a pianist and composer. Active in Paris, he simplified the Erard double escapement action; the Herz-Erard model remained the prototype of all modern grand repetition actions. He also invented a practice device, the 'Dactylion'. With one Klepfer he established a factory in 1825 and set up on his own in 1851; the business was continued until *c*1930 by Amadée Thibout & Cie.

Hoffmann, Johann Wilhelm (1764–1809). German maker, who took over the business of CHRISTIAN GOTTLOB HUBERT.

Hofmann, Ferdinand (*c*1756–1829). Austrian maker. In 1808 he was chairman of the 'Bürgerlich' organ and piano makers in Vienna and was appointed maker to the court in 1812.

Hopkinson. English firm. It was founded in Leeds by 1840 by John Hopkinson, who in 1846 also opened a London factory. The latter eventually merged with Rogers as the Vincent Piano Co. The firm marketed competitively priced grands and uprights of serviceable quality. In 1963 H. B. Lowry and I. D. Zender took over; the Hopkinson name is still maintained.

Hornung & Moeller. Danish firm. It was founded in Copenhagen in 1827 and in 1842 introduced cast-iron frames for its pianos. Makers to the Danish court, the firm has continued to manufacture.

Hubert, Christian Gottlob (1714–93). German maker of Polish origin. He worked in Ansbach, exporting his reasonably priced pianos to France, England and the Netherlands; these were renowned for their durability and pleasant tone. He was also one of the best-known clavichord makers of his time.

Hupfeld. German firm. Active in Leipzig, it is noted for having introduced various types of player piano in the early 20th century, including the 'DEA' (1905).

Ibach. German firm. It was founded in Beyenburg in 1794 and later moved to Unterbarmen. In 1885 an additional factory was opened solely for uprights. Praised by such composers as Bartók, Webern and Richard Strauss, the Ibach is a solidly crafted instrument. The firm also built some organs.

Janko, Paul von (1856–1919). Hungarian inventor. The Janko Keyboard, patented in 1882, had six tiers; each of the levers on the two interlocking manuals had three touch-points. It enabled all scales to be fingered alike and reduced the octave span from 18·5 cm to 13 cm. It also compensated for the unequal finger-lengths. A number of makers in Europe and the USA briefly adopted it.

Karn. Canadian firm. It made pianos from the late 1880s in Woodstock, Ontario, and player pianos from 1901. At one time Karn also had branches in London and Hamburg; the firm ceased trading in 1920 but Karn pianos were made by Sherlock-Manning up to 1957.

Kemble. English firm. It was founded in London in 1911 by Michael Kemble, who was joined by Victor Jacobs in 1918. By 1933 the partnership had acquired Moore & Moore, B. Squire & Son and Squire & Longson. The 'Jubilee', 'Minx Miniature' and 'Cubist' models all proved popular. The firm took over Cramer (1964) and Brinsmead (1967) and is an agent for Yamaha; it also makes pianos for Chappell.

Kawai. Japanese firm. Established in 1925 and based in Hamamatsu, it is the country's second largest maker after Yamaha. Aside from grands and uprights of superior quality, Kawai also markets electronic keyboards.

Kimball. American firm. It was founded in Chicago in 1857, at first acting solely as a piano dealer. In 1887 it opened a factory and patented improvements to piano plates; the firm's priorities were turnover, quality of construction and competitive pricing. It also made organs. After 1945 pianos became Kimball's principal product. In 1966 the company (part of the Jasper Corporation since 1959) took over Bösendorfer.

Kirckman [Kirkman]. English firm. Chiefly renowned for its harpsichords, the firm also made pianos from *c*1774. Joseph Kirckman (*c*1790–1877) expanded the piano business and a number of patents were taken out, including one in 1870 for the use of wrought steel tension bars and a wrest plank in grands. The firm adopted German-style overstringing early on, but later abandoned it. Kirckman grands featured the Herz-Erard action. In 1896 the firm was absorbed by Collard & Collard.

Klemm & Brother(s). German importers, active in Philadelphia from 1819. John G. Klemm established a piano warehouse and began to sell pianos in 1825. Babcock worked for Klemm from *c*1830 to 1832. The firm appears to have ceased trading after 1879.

Knabe. American firm. It was founded in Baltimore in 1839 by William Knabe (1803–64), a German immigrant, in partnership with Henry Gaehle. On Gaehle's death in 1855 Knabe continued independently, and by 1860 controlled the market in most of the southern USA. His sons expanded the business and Wm. Knabe & Co. became one of the leading American firms, with an annual output of *c*2000 pianos in 1900. Knabe was bought by the American Piano Co. in 1908 and in 1932 became part of the Aeolian group.

Knight, Alfred (1898–1974). English maker. He trained and worked in English firms and founded his own business in 1935. During the 1950s he led the British industry, achieving an annual output of almost 2000. His instruments, though modest, are noted for their sound design and high quality workmanship, and have won international recognition.

Könnicke, Johann Jakob (1756–1811). Austrian maker. He was appointed 'Bürgerlich Instrumentmacher' in Vienna. In 1795 he built a six-manual piano with a six-octave compass ('Piano Forte pour la parfaite Harmonie') to the designs of the Kapellmeister of Linz cathedral, Johann Georg Roser. Haydn and Beethoven played this instrument in 1796.

Kriegelstein. French firm. It was founded in Paris in 1831. As well as making square pianos it introduced in 1842 a small upright, the 'Mignon Pianino', with an unusually rich tone and even registers, and also a new simplified repetition action for the grand. The firm survived until the 1930s.

Lambertini. Portuguese firm. It was founded in Lisbon in the mid-1830s by an Italian and closed in 1922.

Lemme, Friedrich Carl Wilhelm (1747–1808). German maker. He was an organist in Brunswick before establishing a workshop with his father, a clavichord and piano maker. By 1787 they had made 800 instruments. He built large grand pianos of five and a half octaves in the English style as well as squares; Forkel (1782) described them as 'among the best, both as regards workmanship and tone'.

Lichtenthal, Herman (*fl c*1830–51). Active in Brussels, he is best known for inventing an upright piano action (patented 1832) that used a leather thong, anticipating the tape-check mechanism of the modern instrument.

Lipp. German firm of instrument makers. It was founded *c*1875 in Stuttgart and from 1895 to 1965 made mainly pianos. In the 1950s it introduced a number of electronic instruments, including several devices of the piano attachment type.

Logier, Johann Bernhard (1777–1846). German pianist and inventor. Working in Dublin in 1814 he invented the 'Chiroplast', a mechanism which guided both wrist and finger position with the use of a frame and rails placed over the length of the keyboard. The device was sold as late as 1877.

Longman & Broderip. English firm. It was founded *c*1767 by James Longman and others in London and was known as Longman, Lukey & Co., 1769–75. Francis Broderip joined in 1775 and Lukey withdrew in 1776. It marketed a variety of keyboard instruments, including pianos housed in furniture items. After bankruptcy in 1798, Longman went into partnership with CLEMENTI.

Loud. Anglo-American family of makers. Thomas Loud (*c*1765–1833) patented a small upright (London, 1802) with diagonal stringing; he went to New York, *c*1816, and was building overstrung 'piccolo' uprights by 1830. He may or may not be connected with the Philadelphia firm of Loud & Brothers, which from the late 1820s to the 1860s patented various devices and was highly prolific (exporting to the West Indies and South America).

Mackay. American mercantile family. John Mackay (*d* 1841) financed such Boston makers as Babcock and Chickering, entering into partnership with the latter from 1830; he also imported woods and other materials. His nephew George D. Mackay (*d* 1824) set up a factory by 1823 with Babcock as superintendent. John's mother, Ruth (1743–1833), also supported Babcock.

Maendler–Schramm. German firm. It was founded about 1907 by Karl Maendler, who had taken over the Munich piano factory of his father-in-law, M. J. Schramm. The main output was of heavily built, mass-produced instruments. The business passed to Ernst Zucker in 1956.

Malmsjö. Swedish firm. It was established in Gothenburg in 1843 by Johan Gustav Malmsjö (1815–91) and by the end of the century was the only maker of grands in Sweden. The firm ceased production in 1977.

Mangeot, Edouard J. (1835–98). French maker. In 1859 he took over his father's business, exhibiting in London in 1862. He visited the USA in 1866 and became a firm advocate of the American style of piano building, winning a gold medal at the 1878 Paris exhibition. But about 1888 his Paris business failed, owing in part to French conservatism. In 1889 he founded the journal *Le monde musical*, in which he championed the American system of piano making.

Mason & Hamlin. American firm. Founded in Boston, it made reed organs from 1854 and pianos from 1883. In 1900 it patented the 'tension resonator', which stabilized the soundboard, thus counteracting loss of tone, and also the 'duplex scale', which used sympathetic strings to increase sonority. From *c*1900 to the 1920s the firm was among the foremost American makers, producing a small number of quality grands and uprights. It later was taken over by the Aeolian American Corporation.

Meacham. American family, making wind instruments and also pianos prior to *c*1850.

Merlin, John Joseph (1735–1803). Flemish maker and inventor. He came to London in 1760 and in 1774 patented a combined harpsichord and piano, with a down-striking action. Such an instrument of 1780 is also fitted with an apparatus for recording a performance on a paper band. In 1777 he made a six-octave grand for Charles Burney, especially for duets, 'the first that was ever constructed'.

Meyer, Conrad (*d* 1881). American maker of German birth. He settled in Baltimore in 1819 and in 1829 started his own firm in Philadelphia. He claimed to have invented the single cast-iron frame, but the credit for this belongs to Babcock. His pianos were said to be of excellent quality. The firm continued into the 1890s.

Moór, Emanuel (1863–1931). Hungarian composer and pianist. He invented the piano bearing his name – a double-manual instrument with an upper keyboard that can be played separately or coupled for octave doubling.

Mooser, (Jean Pierre Joseph) Aloys (1770–1839). Swiss maker. Active in Fribourg, he also made organs. Erard of Paris attempted to enlist his skill. His sons continued the firm but failed to maintain its reputation.

Müller, Matthias (1769–1844). German maker. In 1800 in Vienna he invented an upright piano, brought out under the name 'Ditanaklasis'; its action was an adaptation of the Viennese grand action. He also invented the 'Gabel-Harmon-Pianoforte' (1827), a single-strung instrument with hoop-like tuned tuning-forks designed to give additional resonance and aid tuning. (*See also* JOHN ISAAC HAWKINS.)

Neupert. German firm. It was founded in Nurenburg in 1868. In 1907–8 it was among the first German firms to produce fortepianos modelled closely on 18th-century prototypes, as well as other early keyboard instruments.

Nunns & Clark. American firm. Active in New York, 1836–60, it was founded by R. and W. Nunns, who had started manufacturing in 1823. They were joined by the Englishman John Clark in 1833 and made some highly ornate square pianos with a conventional action.

Osborne, John (*c*1792–1835). American maker. Apprenticed to Crehore, *c*1808–14, he founded his own firm in Boston by 1815, building well-crafted grand, upright, square and cabinet pianos. He also worked briefly with Stewart and for Meacham & Pond. In 1833 he moved to New York.

Packard. American firm. Originally a maker of reed organs, from 1893 it also (and later chiefly) made pianos renowned for their deluxe cabinet work.

Pape, Jean Henri (1789–1875). French maker of German birth. He founded his own business in Paris in 1815, after helping Pleyel. He took out 137 patents, but many were never widely adopted. However, his down-striking French grand action was adopted throughout Europe and America; he also experimented with felt-covered hammers and used tempered steel wire for strings. In 1828 he patented the 'pianino', an upright a metre high with the earliest use of overstringing, which had enormous vogue in France and England.

Petzoldt [Petzold], **Wilhelm Lebrecht** (1784 after 1829). German maker. He was apprenticed to Walter and then moved to Paris. One of the most innovatory makers of his day, he made pianos with an English action, worked on an improved escapement, used longer and heavier strings for greater volume and (in 1829) introduced a square piano with a cast-iron frame, open at the sides and base, leaving the soundboard independent of this structure.

Pfeiffer. German family of makers. The firm was established in Stuttgart, *c*1862 by Joseph Anton Pfeiffer (1828–81). His son Carl Anton (1861–1927) studied in Steinway factories and elsewhere, and subsequently developed pedal pianos for attachment to uprights and grands, transposition mechanisms and tools for piano making. The firm has remained in the family; its output is small, but of the highest quality.

Pleyel. French firm. Established by Ignace Pleyel in Paris, 1807, it developed the best features of English pianos and built 'pianinos' (1811–15) with the help of Pape. Chopin, who admired the Pleyel's subtle sonority and light touch, owned a grand of 1839. Auguste Wolff (1821–87) joined in 1855, the firm becoming Pleyel, Wolff & Cie. On Wolff's death Gustave Lyon (1857–1936) took control; Pleyel, Lyon & Cie also made practice keyboards, two-manual pianos and the 'Pleyela', a reproducing player mechanism. A high level of production was achieved: by the 1870s annual output had reached 2500. In 1961 Pleyel merged with Erard-Gaveau, which was bought by Schimmel in 1971.

Pohlmann, Johannes (*fl* 1767–93). English maker of German origin. He was probably the second piano maker in London after Zumpe, fulfilling some of the latter's orders and active from *c*1767. His pianos, possibly all squares, had a range of five octaves or less, two hand stops, and the English action with overdampers.

Rachals. German firm. The founder was Matthias F. Rachals (1801–66), who began manufacture in Hamburg in 1832. By 1845 the firm had sold over 1000 instruments, specializing in pianos that could easily be dismantled for transportation.

Rogers. English firm. It was founded in London in 1843 by George Rogers. Shortly after 1918 it merged with Hopkinson, and in 1963 the two were taken over by H. B. Lowry and I. D. Zender, who redesigned the scaling and casework. Production of grands ceased and the firm now concentrates on durable, quality uprights.

Rolfe. English firm. Also music publishers and sellers, the firm was founded *c*1785 in London. It patented the earliest specification for 'Turkish music', where a hammer strikes the soundboard to produce the sound of a drum, and manufactured pianos until 1888.

Roller, Jean [Johann]. German maker. He worked in Paris from *c*1808 and later became a partner in the French firm of BLANCHET, retiring in 1851.

Rosenkranz. German firm. It was founded in Dresden in 1797 by Ernst Rosenkranz and manufactured grands, small uprights, giraffe pianos (see fig.15) and, for a time, Janko keyboards.

Schantz, Johann (1762–1828). Austrian maker. Haydn admired the pianos made by Johann's brother Wenzel, whose workshop Johann took over in 1791 after Wenzel's death. In 1796 he was described as a master second only to Anton Walter.

Scherr, Emilius Nicolai (1794–1874). American maker of Danish birth. He trained in Europe, arriving in Philadelphia with C. F. L. Albrecht in 1822, and exhibited a square piano in 1832. His instruments are rare and beautifully constructed.

Schiedmayer. The name of two German firms. The first was established in 1809 by Johann Lorenz Schiedmayer, who came from an established family of makers including his father Johann David (1753–1805), and Carl Dieudonné in Stuttgart. It began making uprights in 1842 and was successful with concert and domestic instruments. The second firm, founded in 1853 by Schiedmayer's younger sons, Julius and Paul, began making pianos from *c*1860, with competition between the two producing quality instruments, although the second grew more rapidly in reputation. The two merged in 1969; their combined output numbered *c*126,700 by 1980.

Schimmel. German firm. It was founded in 1885 in Leipzig. In 1930 it moved to Brunswick and after 1945 expanded greatly, in 1980 producing 29% of the total West German output. In 1971 Schimmel took over the Pleyel group and has continued to produce Pleyel, Erard and Gaveau instruments.

Schmahl. Swabian family of instrument makers. In 1774 Christoph Friedrich Schmahl entered into partnership with his father-in-law FRANZ JAKOB SPÄTH, who invented the 'Tangentenflügel'. The firm, based in Regensburg, produced such instruments from *c*1790 and also gained a considerable reputation for fortepianos with Viennese action.

Schmidt-Flohr. Swiss company. Founded in Berne in 1833, it made quality pianos and in 1921 produced a double keyboard that was the prototype for Emanuel Moór's instrument. Its pianos have been made by Kemble since 1977.

Schnell, Johann Jakob (*b* 1740). German maker. He worked in Rothenburg, then in the Netherlands, Paris and Ludwigsburg. As court maker to the Duchess of Artois, he built an 'anémocorde' ('wind-piano') in 1789. He also made harpsichords and organs.

Schröter, Christoph Gottlieb (1699–1782). German designer. Independently of Cristofori he drew up designs for a piano action in 1717; no instruments using his action have survived.

Schweighofer. Austrian firm. It was founded in Vienna *c*1792. Johann Michael Schweighofer (1806–52), son of the founder, started his own firm, concentrating on well-crafted quality instruments. It ceased production in 1938.

Silbermann. German family of makers. Although the family was renowned chiefly as organ builders, Gottfried Silbermann (1683–1753) was making grand pianos by 1736 based on the Cristofori action, but with a heavier case, framing and stringing. J. S. Bach showed interest in such instruments, and Silbermann supplied a number to Frederick the Great. Two nephews, Johann Heinrich and Johann Daniel Silbermann, also made grands (see fig.3).

Sohmer, Hugo (1846–1913). American maker of German birth. In New York from 1863, he founded a company in partnership with K. Kuder in 1872. It expanded rapidly after 1876. The firm was owned by the Sohmer family until 1983 and is still active.

Southwell, William (1756–1842). Irish maker. He worked mostly in Dublin, but patented influential designs in London: his 1794 patent extended the compass of the square piano; that of 1798 placed the piano on its side to make a small upright and used a 'sticker action' (see p.28); the 1807 patent was for an improved cabinet piano; and the 1811 patent was for an upright with a soundboard and strings sloping away from the player. Few of his instruments have survived.

Späth, Franz Jakob (1714–86). German maker. He invented the 'Tangentenflügel', in which freely moving slips of wood, like harpsichord jacks, replaced the hammers. In 1774 he founded a firm in partnership with C. F. SCHMAHL. He also built organs.

Spencer. English firm. It was established in London in 1884. Production concentrated on cheap, durable uprights, with few grands. The business closed during World War II.

Steck. American firm. Founded in New York by George Steck in 1857, it won many awards at European exhibitions in the 1870s and opened a factory in Gotha, Germany. In 1904 it was taken over by the Aeolian Co.

Stein, Johann (Georg) Andreas (1728–92). German maker. He was a noted organ builder, working chiefly in Augsburg. From the 1760s he concentrated on other keyboard instruments, experimenting with various 'combination' instruments, including the 'Poli-Toni-Clavichordium' (a harpsichord-*cum*-piano), 'Melodika' (organ-*cum*-keyboard instrument), 'vis-à-vis Flügel' (similar to the Poli-Toni-Clavichordium), 'clavecin organisé' (organ-*cum*-piano) and the 'Saitenharmonika' (piano-*cum*-spinet). Stein devised the type of piano and action which when copied in Vienna established the city's piano-making pre-eminence; he is often credited with inventing the Prellmechanik in which the Prelleiste is replaced with individually hinged and spring-loaded escapement levers (see pp.12–18). His children Nannette (1769–1833) and Matthäus Andreas (1776–1842) also became noted makers; active in Vienna, they were partners until 1802, when Matthäus Andreas established his own firm, André Stein, which was taken over by his son Carl Andreas on his death.

Steingräber. German firm. It was founded by Eduard Steingräber (1823–1906) in 1852 in Bayreuth, where his descendants continue to make pianos.

Steinway. American firm. It was founded in 1853 by Heinrich Engelhard Steinweg (1797–1871), a German builder who had emigrated to New York in 1850. Steinway & Sons began by producing overstrung iron-framed square pianos and in 1859 Heinrich's son Henry (1831–65) patented a grand similarly overstrung. After the deaths of Henry and his brother Charles, C(arl) F(riedrich) Theodor Steinweg emigrated to New York, having sold the family firm in Germany (*see* GROTRIAN-STEINWEG). By 1880, the year in which Steinway opened a factory in Hamburg, Theodore Steinway (as he became in the USA) had developed the modern grand, with its massive iron frame overstrung with heavy strings at high tension (see figs.28 and 29); annual production at this time was about 2500 pianos. He registered patents for the duplex scale, the cupola metal frame, the capo d'astro agraffe and also 38 others. His brother Albert (1840–77) patented improvements to the sostenuto pedal in 1874 (for square pianos) and 1875 (grands). Another brother William (1836–96) was responsible for the commercial and promotional development of the company into a world-wide venture. The firm soon established an almost unrivalled reputation for pianos with a sonorous and dynamic tone and high consistency of quality; since World War II Steinways have dominated concert platforms. In 1972 Steinway was bought by CBS.

Stewart, James (*b* late 1700s; *d* ?after 1860). British maker. In 1812 he went to Baltimore and after working in Philadelphia and with Osborne in Boston founded the firm of Stewart & CHICKERING in 1823. Back in London in 1826 he was foreman of Collard & Collard for 35 years. He issued seven patents, one of which formed the basis of modern stringing methods by using a continuous wire of double length and a single hitch-pin instead of the usual two strings. His instruments were of fine craftsmanship, praised as 'unrivalled in tone, touch, and action'.

Stieff. American firm. It was founded by the German Charles Maximilian Stieff in Baltimore during the 1850s. His family continued the business until it failed in 1952.

Stodart. English firm. It was founded in 1775 in London by Robert Stodart (1748–1831), who had worked with Broadwood and Backers on the invention of the English grand action. Ownership passed to his nephews Matthew and William in 1792; it was in the Stodart workshop that Allen and Thom invented the 'compensation' frame (see p.41). On the death of Malcolm Stodart (William's son) in 1861 the firm ceased production.

Streicher. Austrian firm. It was founded in 1802 by the pianist Nannette Streicher (1769–1833), daughter of Johann Andreas Stein, and became the most eminent firm in Vienna. Streicher perfected the Viennese action, although it built Anglo-German- and English-action pianos when the popularity of the former declined. The firm remained in the family until it ceased trading in 1896 (see pp.33–4). Surviving grands are beautifully veneered and usually have four pedals: una corda, bassoon, pianissimo and damper.

Taskin, Pascal (Joseph) (1723–93). French maker. Although best known as a harpsichord maker, he increasingly turned to building grand pianos from the late 1770s. His instruments, though finely crafted and often beautifully veneered, had rather primitive actions with no escapement, but they worked well.

Thom, James. Co-inventor with WILLIAM ALLEN of the 'compensation' frame.

Tischner, A. (*fl c*1800–30). Maker probably of German origin, active in St Petersburg. His grands, with a range of over six octaves, resembled contemporary English instruments and had high quality cabinet work.

Tomkison, Thomas (*fl* 1798–1851). English maker. He made fine quality grand and square pianos, working in London. His later square pianos have metal bracing, a compass of six octaves and use a heavier case.

Virgil, A(lmon) K(incaid) (1839 or 1842–1921). American inventor. In New York, *c*1872, he devised a silent practice keyboard, which was subsequently widely marketed; he formed the Virgil Practice Clavier Co. in 1890.

Wagner, Johann Gottlob (1741–89). German maker. A pupil of the Silbermanns, he worked in Dresden and in 1774 invented the 'clavecin royal', a square piano of about five octaves, modified from Cristofori's instrument, but with uncovered wooden hammers.

Walter, (Gabriel) Anton (1752–1826). Austrian maker of German birth. He worked in Vienna from the 1780s. His pianos, noted for the sophistication of their actions, are considered the foremost of their period and were favoured by Mozart in his later years (see also pp.18 and 29–33).

Weber. American firm. It was founded in 1852 in New York by Albert Weber (1828–79), who had emigrated to the USA from Germany. Although not innovatory, Weber's pianos were of a high quality and were used by such pianists as Hofmann and Paderewski. Among the largest producers in the USA, the firm was bought by the Aeolian Co. in 1903, the name being retained.

Welte. German firm. It specialized in mechanical musical instruments; in 1904 Edwin Welte devised the 'Welte-Mignon' PLAYER PIANO, which was pre-eminent in its field. The firm, based in Freiburg, closed *c*1944.

Wornum. English firm of instrument makers. The piano business was founded in the 1770s in London by Robert Wornum (1780–1852). He took out patents for diagonally and vertically strung low upright pianos (1811, 1813), and was important in developing the industry for small pianos for home use. By 1828 he completed the development of his cottage piano action, and in 1842 patented the tape-check action for uprights that facilitated rapid repetition (a device still in use). (See also p.44 and fig.23.)

Wurlitzer. American firm. Founded in Cincinnati by Franz Rudolph Wurlitzer (1831–1914), a German immigrant, it became best known for its military and mechanical instruments, although pianos with the name were made from 1880. In 1985 it acquired certain assets formerly owned by Aeolian Pianos, Inc., including the trade names Cable, Chickering, J. & C. Fischer, Mason & Risch, Henry F. Miller and Winter.

Yamaha. The brand name of musical instruments (also of motorcycles etc.) made by the Japanese firm Nippon Gakki, founded in 1887 in Tokyo by Torkusu Yamaha (1851–1916). The first upright pianos were produced in 1900 and small grands from 1910. In the early years advice was sought from Bechstein. The firm's concert grands are of fine quality, and have been used by artists such as Richter. The consistently fine uprights (including the U1H model) are among the most popular today. (See also p.68.)

Zumpe, Johannes (*fl* 1735–83). English maker of German origin. He probably worked for the Silbermanns and briefly for Shudi before setting up his own workshop in 1761 in London. His earliest square pianos were based on those of Cristofori, though simpler, with the English single action (see p.21). In the 1780s Zumpe designed a double action without escapement (see p.25). A typical Zumpe square had a small soundboard, light hammers and two hand stops; the dynamic range was very limited.

Bibliography

HISTORY OF THE PIANO

Basic literature

D. Spillane: *History of the American Pianoforte* (New York, 1890/*R*1969)

R. E. M. Harding: *The Piano-forte: its History Traced to the Great Exhibition of 1851* (Cambridge, 1933, rev. 2/1978)

E. Closson: *Histoire du piano* (Brussels, 1944; Eng. trans. as *History of the Piano*, 1947, rev. 2/1974)

A. Loesser: *Men, Women and Pianos: a Social History* (New York, 1954)

H. R. Hollis: *The Piano: a Pictorial Account of its Ancestry and Development* (Newton Abbot and London, 1975)

D. Wainwright: *The Piano Makers* (London, 1975)

C. Ehrlich: *The Piano: a History* (London, 1976)

M. Bilson: 'The Viennese Fortepiano of the Late 18th Century', *Early Music*, viii (1980), 158

E. M. Good: *Giraffes, Black Dragons, and other Pianos: a Technological History from Cristofori to the Modern Concert Grand* (Stanford, Calif., 1982)

H. Schott: 'From Harpsichord to Pianoforte: a Chronology and Commentary', *Early Music*, xiii (1985), 28

General reference

BurneyH

S. Maffei: 'Nuova invenzione d'un gravicembalo col piano, e forte, aggiunte alcune considerazioni sopra gl'instrumenti musicale', *Giornale dei letterati d'Italia*, v (Venice, 1711), 144; Ger. trans. in J. Mattheson: *Critica musica* (Hamburg, 1722–5/*R*1964), 335

C. P. E. Bach: *Versuch über die wahre Art das Clavier zu spielen* (Berlin, 1753–62, 2/1787–97/*R*1957; Eng. trans., 1949)

Ancelet: *Observations sur la musique, les musiciens et les instruments* (Amsterdam, 1757)

C. Burney: *The Present State of Music in France and Italy* (London, 1771, 2/1773); ed. P. Scholes as *Dr. Burney's Musical Tours* (London, 1959)

——: *The Present State of Music in Germany, the Netherlands and United Provinces* (London, 1773, 2/1775); ed. P. Scholes as *Dr. Burney's Musical Tours* (London, 1959)

'Ditanaklasis', *AMZ*, vi (1803–4), 367

K. C. Krause: 'Nachricht über eine wesentliche Verbesserung der Klaviature der Tasteninstrumente', *AMZ*, xii (1809–10), 649

C. F. G. Thon: *Über Klavierinstrumente: deren Ankauf, Behandlung und Stimmung* (Sondershausen, 1817)

C. Kützing: *Das Wissenschaftliche der Fortepiano-Baukunst* (Berne, 1844)

W. Pole: *Musical Instruments in the Great Industrial Exhibition of 1851* (London, 1851)

'Reports on Musical Instruments', *Exhibition of the Works of Industry of All Nations 1851: Reports of the Juries*, ii (London, 1852), 705

J. Fischhof: *Versuch einer Geschichte des Clavierbaues* (Vienna, 1853)

C. Schafhäutl: *Die Pianofortebaukunst der Deutschen* (Berlin, 1854)

K. A. André: *Der Klavierbau* (Frankfurt am Main, 1855)

H. Welcker von Gontershausen: *Der Flügel* (Frankfurt am Main, 1856)

E. F. Rimbault: *The Pianoforte: its Origins, Progress and Construction, with some account of the Clavichord, Virginal, Spinet, Harpsichord etc.* (London, 1860)

E. Brinsmead: *The History of the Pianoforte* (London, 1868, 4/1889)

A. de Pontécoulant: *La musique à l'exposition universelle* (Paris, 1868)

F. J. Fétis: 'Instruments de musique', *Exposition universelle de 1867 à Paris: rapports du jury international*, ed. M. .M. Chevalier, ii/10 (Paris, 1868), 237

O. Paul: *Geschichte des Klaviers* (Leipzig, 1868)

G. F. Sievers: *Il pianoforte* (Naples, 1868)

O. Comettant: *La musique, les musiciens, et les instruments de musique* (Paris, 1869)

P. Stevens: 'Report upon Musical Instruments', *Paris Universal Exposition 1867: Reports of the U.S. Commissioners* (Washington, DC, 1869)

H. Welcker von Gontershausen: *Der Klavierbau* (Frankfurt am Main, 1870)

E. Schelle: 'Musikalische Instrumente', *Officieller Ausstellungs-Bericht*, xv (Vienna, 1873) [report of Vienna Exhibition of 1873]

C. Ponsicchi: *Il pianoforte: sua origine e sviluppo* (Florence, n.d.)

A. Marmontel: *Histoire du piano et de ses origines* (Paris, 1885)

E. de Briqueville: *Le piano de Mme. DuBarry et le clavecin de la reine Marie-Antoinette* (Versailles, 1892)

F. M. Smith: *A Noble Art: Three Lectures on the Evolution and Construction of the Piano* (New York, 1892)

E. de Briqueville: 'Le piano à Versailles sous Marie-Antoinette', *Revue de l'histoire de Versailles et de Seine-et-Oise*, vii (1906), 193

——: *Les ventes d'instruments de musique au XVIIIe siècle* (Paris, 1908)

J. Blüthner and H. Gretschel: *Der Pianofortebau* (Berlin, 1909)

G. Kinsky: *Katalog des Musikhistorischen Museums von Wilhelm Heyer in Cöln*, i (Leipzig, 1910)

H. E. Krehbiel: *The Pianoforte and its Music* (London and New York, 1911)

PIANO

A. Kraus: 'Italian Inventions for Instruments with a Keyboard', *SIMG*, xiii (1911–12), 441

E. J. Dent: 'The Pianoforte and its Influence on Modern Music', *MQ*, ii (1916), 271

S. Wolfenden: *A Treatise on the Art of Pianoforte Construction* (London, 1916–27/*R*1975)

R. S. Clay: 'The British Pianoforte Industry', *Journal of the Royal Society of Arts*, lxvi (1918), 154

W. Pfeiffer: *Taste und Hebeglied des Klaviers: eine Untersuchung ihrer Beziehungen im unmittelbaren Angriff* (Leipzig, 1920; Eng. trans., 1967)

M. de Guchtenaere: *Le piano: son origine, sa facture* (Paris, 1926)

H. Neupert: *Vom Musikstab zum modernen Klavier* (Bamberg, 1926)

L. Nalder: *The Modern Piano* (London, 1927)

E. Blom: *The Romance of the Piano* (London, 1928)

M. A. Blondel: 'Le piano et sa facture', *EMDC*, II/iv (1929), 2061

P. James: *Early Keyboard Instruments from their Beginnings to the Year 1820* (London, 1930/*R*1967)

H. Brunner: *Das Klavierklangideal Mozarts und die Klaviere seiner Zeit* (Augsburg, 1933)

A. Casella: *Il pianoforte* (Milan, 1937)

A. E. Wier: *The Piano* (New York, 1940)

C. Parish: 'Criticisms of the Piano when it was New', *MQ*, xxx (1944), 428

W. Pfeiffer: *Vom Hammer: Untersuchungen aus einem Teilgebiet des Flügel- und Klavierbaus* (Stuttgart, 1948/*R*1962; Eng. trans., 1978)

H. Freygang: *Die Produktions- und Absatzbedingungen der deutschen Klavierindustrie* (diss., Humboldt U., Berlin, 1949)

N. E. Michel: *Michel's Piano Atlas* (Rivera, Calif., 1953)

V. Luithlen: 'Der Eisenstädter Walterflügel', *MJb 1954*, 206

F. J. Hirt: *Meisterwerke des Klavierbaues* (Olten, 1955; Eng. trans., 1968)

U. Rück: 'Mozarts Hammerflügel erbaute Anton Walter Wien', *MJb 1955*, 246

D. H. Boalch: *Makers of the Harpsichord and Clavichord* (London, 1956, rev. 2/1974)

V. A. Brady: *Music for the Millions: the Kimball Piano and Organ Story* (Chicago, 1957)

C. Clutton: 'The Pianoforte', *Musical Instruments through the Ages*, ed. A. Baines (London, 1961), 68–102

W. Pfeiffer: *Über Dämpfer, Federn und Spielart* (Frankfurt am Main, 1963)

P. Locard and R. Stricker: *Le piano* (Paris, 1966)

W. L. Sumner: *The Pianoforte* (London, 1966)

Der klangliche Aspekt beim Restaurieren von Saitenklavieren: Graz 1971

O. Rindlisbacher: *Der Klavierbau in der Schweiz* (Berne and Munich, 1972)

C. F. Colt: 'Early Pianos, their History and Character', *Early Music*, i (1973), 27

C. I. Walsh: *An Economic and Social History of the Pianoforte in Mid- and Late-Victorian Britain* (diss., U. of London, 1973)

D. S. Grover: *The Piano: its Story from Zither to Grand* (London, 1976)
C. F. Colt and A. Miall: *The Early Piano* (London, 1981)
D. Gill, ed.: *The Book of the Piano* (Oxford, 1981)
S. K. Taylor: *The Musician's Piano Atlas* (London, 1981)
F. Schulz: *Pianographie* (Recklinghausen, 1982)
M. Bilson: 'Beethoven and the Piano', *Clavier*, xxii/8 (1983), 18
O. Barli: *La facture française du piano de 1849 à nos jours* (Paris, 1983)
V. Vitale: *Il pianoforte a Napoli nell' ottocento* (Naples, 1983)
K. Ford: 'The Pedal Piano – a New Look', *The Diapason*, lxxv (1984) no.10, p.10; no.11, p.6; no.12, p.14
R. Burnett: 'English Pianos at Finchcocks'; *Early Music*, xiii (1985), 45
V. Pleasants: 'The Early Piano in Britain (*c*1760–1800)', *Early Music*, xiii (1985), 39
S. Pollens: 'The Early Portuguese Piano', *Early Music*, xiii (1985), 18
H. Schott: *Catalogue of Musical Instruments in the Victoria and Albert Museum*, i: *Keyboard Instruments* (London, 1985)
W. H. Cole: 'The Early Piano in Britain Reconsidered', *Early Music*, xiv (1986), 563
M. Hood: 'Nannette Streicher and her Pianos', *Continuo*, x (1986), no.5, p.2; no.6, p.2

Specific studies

A. J. Hipkins: *A Description and History of the Pianoforte* (London, 1896, rev. 3/1929/*R*1975)
W. B. White: *Theory and Practice of Pianoforte Building* (New York, 1906)
D. Spillane: *The Piano: Scientific, Technical and Practical Instructions relating to Tuning, Regulating and Toning* (New York, 1907)
W. H. G. Flood: 'Dublin Harpsichord and Pianoforte Makers of the Eighteenth Century', *Journal of the Royal Society of Antiquaries of Ireland*, xxxix (1909), 137
A. Dolge: *Pianos and their Makers* (Covina, Calif., 1911/*R*1972)
W. B. White: *The Player-piano Up-to-date* (New York, 1914)
T. Cieplik: *Entwicklung der deutschen Klavierindustrie bis zu ihrer heutigen Bedeutung als Exportindustrie* (diss., U. of Giessen, 1923)
C. Sachs: *Das Klavier* (Berlin, 1923)
G. Roos: *Die Entwicklung der deutschen Klavierindustrie* (diss., Humboldt U., Berlin, 1924)
N. Broder: 'Mozart and the Clavier', *MQ*, xxvii (1941), 422
H. Gough: 'The Classical Grand Pianoforte', *PRMA*, lxxvii (1950–51), 41
H. Haupt: 'Wiener Instrumentenbauer von 1751 bis 1815', *SMw*, xxiv (1960), 120–84
R. Benton: 'The Early Piano in the United States', *HMYB*, xi (1961), 179
E. D. Blackham: 'The Physics of the Piano', *Scientific American*, ccxiii (1965), 88
W. S. Newman: 'Beethoven's Piano versus his Piano Ideals', *JAMS*, xxiii (1970), 484

PIANO

A. Ord-Hume: *Player Piano* (London, 1970)
D. Melville: 'Beethoven's Pianos', *The Beethoven Companion*, ed. D. Arnold and N. Fortune (London, 1971), 41
M. Bilson: 'Schubert's Piano Music and the Pianos of his Time', *Piano Quarterly*, xxvii (1978–9), 56
E. Badura-Skoda: 'Prologomena to a History of the Viennese Fortepiano', *Israel Studies in Musicology*, ii (1980), 77
D. Wainwright: *Broadwood by Appointment: a History* (London, 1983)
R. Winter: 'The Emperor's New Clothes: Nineteenth-century Instruments Revisited', *19th Century Music*, vii (1983–4), 251
S. Pollens: 'The Pianos of Bartolomeo Cristofori', *JAMIS*, x (1984), 32
D. Wythe: 'The Pianos of Conrad Graf', *Early Music*, xii (1984), 447

PIANO PLAYING

C. P. E. Bach: *Versuch über die wahre Art das Clavier zu spielen* (Berlin, 1753–62, 2/1787–97/*R*1957; Eng. trans., 1949)
F. W. Marpurg: *Anleitung zum Clavierspielen* (Berlin, 1755, repr. Amsterdam, 1760/*R*1970, 2/1765)
D. G. Türk: *Clavierschule* (Leipzig and Halle, 1789/*R*1962; Eng. trans., 1982)
J. L. Dussek: *Instructions on the Art of Playing the Piano Forte or Harpsichord* (London, 1796)
M. Clementi: *Introduction to the Art of Playing on the Piano Forte* (London, 1801/*R*1974)
L. Adam: *Méthode de piano* (Paris, 1802, 2/1805/*R*1974)
A. Streicher: *Kurze Bemerkungen über das Spielen, Stimmen und Erhalten der Forte-piano* (Vienna, 1802)
J. N. Hummel: *Ausführlich theoretisch-practische Anweisung zum Piano-forte Spiel* (Vienna, 1828, rev. 2/1838; Eng. trans., 1829)
C. Montal: *L'art d'accorder soi-même son piano* (Paris, 1836/*R*1976)
C. Czerny: *Letters to a Young Lady on the Art of Playing the Pianoforte*, ed. and trans. J. A. Hamilton (New York. ?1837–41)
W. von Lenz: *Die grossen Pianoforte-Virtuosen unserer Zeit aus persönlicher Bekanntschaft* (Berlin, 1872; Eng. trans., 1899)
A. Marmontel: *L'art classique et moderne du piano* (Paris, 1876)
A. Fay: *Music Study in Germany* (Chicago, 1880/*R*1965)
F. Kullak: *Aesthetics of Piano-forte Playing* (New York, 1893)
E. Pauer: *A Dictionary of Pianists and Composers for the Pianoforte* (London, 1895)
C. E. and M. Hallé, eds.: *Life and Letters of Sir Charles Hallé* (London, 1896/*R*1975), 57
M. Jaëll: *Le mécanisme du toucher* (Paris, 1897)
C. Weitzman: *A History of Pianoforte-playing and Pianoforte Literature* (New York, 1897/*R*1969)
F. Kullak: *Beethoven's Piano-playing* (New York, 1901)

M. Brée: *Die Grundlage der Methode Leschetizky* (Mainz, 1902, 4/1914; Eng. trans., 1902)
T. Matthay: *The Act of Touch* (London, 1903)
M. Prentner: *The Modern Pianist/Der moderne Pianist* (London, Philadelphia and Vienna, 1903)
R. M. Breithaupt: *Die natürliche Klaviertechnik* (Leipzig, 1905–6)
J. Hofmann: *Piano Playing* (New York, 1908/*R*1976)
——: *Piano Questions Answered* (New York, 1909/*R*1976)
E. Newcomb: *Leschetizky as I knew him* (New York, 1921/*R*1967)
J. Lhevinne: *Basic Principles in Pianoforte Playing* (Philadelphia, 1924/*R*1972 with preface by R. Lhevinne)
T. P. Fielden: *The Science of Pianoforte Technique* (London, 1927, 2/1934)
A. Cortot: *Principes rationnels de la technique pianistique* (Paris, 1928; Eng. trans., 1937)
O. Ortmann: *The Physiological Mechanics of Piano Technique* (London and New York, 1929)
W. Gieseking and K. Leimer: *Modernes Klavierspiel nach Leimer-Gieseking* (Mainz, 1930, 3/1938 with suppl.; Eng. trans., 1932/*R*1948 as *The Shortest Way to Pianistic Perfection*, *R*1972 with suppl. as *Piano Technique* pts.i and ii)
E. Bodky: *Der Vortrag alter Klaviermusik* (Berlin, 1932)
T. Matthay: *The Visible and Invisible in Pianoforte Technique* (London, 1932, rev. 2/1947/*R*1982)
J. Ching: *Piano Technique: Foundation Principles* (London, 1934)
A. Cortot: *Cours d'interprétation* (Paris, 1937)
E. J. Hipkins: *How Chopin Played* (London, 1937)
J. Ching: *Piano Playing* (London, 1946)
D. Ferguson: *Piano Interpretation: Studies in the Music of Six Great Composers* (New York, 1947)
A. Coviello: *What Matthay Meant* (London, 1948)
A. Foldes: *Keys to the Keyboard* (London, 1950)
L. Bonpensiere: *New Pathways to Piano Technique* (New York, 1953)
E. and P. Badura-Skoda: *Mozart-Interpretation* (Vienna and Stuttgart, 1957; Eng. trans., 1962, as *Interpreting Mozart on the Keyboard*)
J. Gát: *A zongorajáték technikája* (Budapest, 1958; Eng. trans., 1965, as *The Technique of Piano Playing*)
G. G. Neigauz [H. Neuhaus]: *Ob iskusstve fortepiannoy igrī* (Moscow, 1958, 3/1967; Eng. trans., 1973, as *The Art of Piano Playing*)
M. Harrison: 'Boogie Woogie', *Jazz*, ed. N. Hentoff and A. McCarthy (New York, 1959/*R*1975)
P. Badura-Skoda, ed.: *Carl Czerny: Über den richtigen Vortrag der sämtlichen Beethoven'schen Klavierwerke* (Vienna, 1963) [annotated reprints from Czerny's *Erinnerungen an Beethoven* and *Vollständigen theoretisch-practischen Pianoforte-Schule* op.500]
H. C. Schonberg: *The Great Pianists* (London, 1963)
J. F. Mehegan: *Contemporary Styles for the Jazz Pianist* (New York, 1964–70)

PIANO

J. Kaiser: *Grosse Pianisten in unserer Zeit* (Munich, 1965, 2/1972; Eng. trans., 1971)
H. Grundmann and P. Mies: *Studien zum Klavierspiel Beethovens und seiner Zeitgenossen* (Bonn, 1970)
D. Barnett: *The Performance of Music: a Study in Terms of the Pianoforte* (London, 1972)
K. Wolff: *The Teaching of Artur Schnabel* (London, 1972, 2/1979 as *Schnabel's Interpretation of Piano Music*)
R. R. Gerig: *Famous Pianists and their Technique* (Newton Abbot, 1976)
L. Kentner: *Piano* (London, 1976)
U. Molsen: *Die Geschichte des Klavierspiels in historischen Zitaten* (Balingen, 1982)
W. Taylor: *Jazz Piano* (Dubuque, Iowa, 1982)
M. Weiss: *Jazz Styles and Analysis: Piano* (Chicago, *c*1982)
J. Last: *Interpretation in Piano Study* (Oxford, 1983)
H. Neuhaus: *The Art of Piano Playing* (London, 1983)
L. Lyons: *The Great Jazz Pianists* (New York, 1983)
V. Vitale: *Il pianoforte a Napoli nell' ottocento* (Naples, 1983)
P. Badura-Skoda: 'Playing the Early Piano', *Early Music*, xii (1984), 477
R. A. Fuller: 'Andreas Streicher's Notes on the Fortepiano', *Early Music*, xii (1984), 461
P. Loyonnet: *Les gestes et la pensée du pianiste* (Montreal, 1985)
L. Nicholson, C. Kite and M. Tan: 'Playing the Early Piano', *Early Music*, xiii (1985), 52
D. Rowland: 'Early Pianoforte Pedalling', *Early Music*, xiii (1985), 5

PIANISTS

J. B. Labat: *Zimmermann et l'école française de piano* (Paris, 1865)
C. F. Weitzmann: *Der letzte der Virtuosen* (Leipzig, 1868)
W. von Lenz: *Die grossen Pianoforte-Virtuosen unserer Zeit* (Berlin, 1872; Eng. trans., 1899/*R*1973)
A Marmontel: *Les pianistes célèbres: silhouettes et medaillons* (Paris, 1878)
——: *Virtuoses contemporains* (Paris, 1882)
T. Currier: 'Paderewski: an Example for the Student of Piano-playing', *The Musician* (1890), May
H. T. Finck: *Paderewski and his Art* (New York, 1895)
O. Bie: *Das Klavier und seine Meister* (Munich, 1898, 3/1921; rev. and Eng. trans., 1899, as *A History of the Pianoforte and Pianoforte Players*)
H. Lahee: *Famous Pianists of Today and Yesterday* (Boston, 1901)
M. Unschuld von Melasfeld: *Die Hand des Pianisten* (Leipzig, 1901, 2/1903)
M. Bree: *Die Grundlage der Methode Leschetizky* (Mainz, 1902, 4/1914; Eng. trans., 1902)
A. Potocka: *Theodor Leschetizky* (New York, 1903)

BIBLIOGRAPHY

A. Hullah: *Theodor Leschetizky* (London, 1906)
E. Baughan: *Ignaz Jan Paderewski* (London, 1908)
J. Bennett: *Forty Years of Music* (London, 1908), 208
A. Schnabel: 'Theodor Leschetizky', *AMz*, xxxvii (1910), 599
N. Bernstein: *Anton Rubinstein* (Leipzig, 1911)
A. Hervey: [*Anton*] *Rubinstein* (London, 1913, 2/1922)
H. Leichtentritt: *Ferruccio Busoni* (Leipzig, 1916), 17
J. Chantovoine: 'Un pianiste homme d'état', *Revu hebdomadaire*, xxviii (April 1919), 368
W. Niemann: *Meister des Klaviers* (Berlin, 1919, rev. 2/1921)
V. Walter: 'Reminiscences of Anton Rubinstein', *MQ*, v (1919), 10
E. Newcomb: *Leschetizky as I Knew him* (New York, 1921/*R*1967)
F. Martens: *Paderewski* (New York, 1923)
H. Brower: 'Some Causes of Paderewski's Leadership in Piano Music', *The Musician* (1926), Sept
W. Raupp: *Eugene d'Albert* (Leipzig, 1930)
S. Nadel: *Ferruccio Busoni* (Leipzig, 1931), 24
H. Dobiey: *Die Klaviertechnik des jungen Franz Liszts* (diss., U. of Berlin, 1932)
I. Philipp: *La technique de Liszt* (Paris, 1932)
E. Dent: *Ferruccio Busoni: a Biography* (London, 1933), 94
H. Landau: *Die Neuerungen der Klaviertechnik bei Franz Liszt* (diss., U. of Vienna, 1933)
R. Klasinc: *Die konzertante Klaviersatztechnik seit Liszt* (diss., U. of Vienna, 1934)
R. Landau: *Ignace Paderewski, Musician and Statesman* (London and New York, 1934/*R*1976)
C. Phillips: *Paderewski* (New York, 1934/*R*1978)
O. Bennigsen: 'The Brothers Rubinstein and their Circle', *MQ*, xxv (1939), 407
C. Drinker Bowen: *'Free Artist': the Story of Anton and Nicholas Rubinstein* (New York, 1939)
W. Lyle: *Rachmaninoff: a Biography* (London, 1939), 211–43
I. Paderewski and M. Lawton: *The Paderewski Memoirs* (New York, 1938/*R*1980)
G. Guerrini: *Ferruccio Busoni: la vita, la figura, l'opera* (Florence, 1944)
G. Kogan: 'Rakhmaninov – pianist', *Sovetskaya musika sbornik*, iv (1945), 58
K. Huschke: *Hans von Bülow als Klavierpädagoge* (Horb, 1948)
A. Strakacz: *Paderewski as I Knew him* (New Brunswick, NJ, 1949)
G. Woodhouse: 'How Leschetizky Taught', *ML*, xxxv (1954), 220
B. Gavoty: *Great Concert Artists: Alfred Cortot* (Geneva and Monaco, 1955)
B. Gavoty and R. Hauert: *Great Concert Artists: Walter Gieseking* (Geneva, 1955)
S. Bertensson and J. Leyda: *Sergei Rachmaninov: a Lifetime in Music* (New York, 1956/*R*1972), 210
A. Henderson: 'Paderewski as Artist and Teacher', *MT*, xcvii (1956), 411
A. Chasins: *Speaking of Pianists* (New York, 1957, rev. 2/1961/*R*1973)

A. H. Eichmann: *Wilhelm Backhaus*, Die Grossen Interpreten (Ghent, 1957; Eng. trans., 1958)
C. Saerchinger: *Artur Schnabel* (London, 1957)
V. Yu Del'son: *Emil' Gilel's* (Moscow, 1959)
'Lhevinne, Rosina', *CBY 1961*
A. Friedheim: *Life and Liszt* (New York, 1961)
B. Moiseiwitsch: 'Leschetizky', *Sunday Telegraph* (3 Sept 1961)
M. Barzetti: 'Walter Gieseking', *Recorded Sound*, i (1961–2), 168
U. Creighton: 'Reminiscences of Busoni', *Recorded Sound*, i (1961–2), 249
H. Ferguson: 'Harold Samuel', *Recorded Sound*, i (1961–2), 186
E. Fisher: 'Busoni and Philipp', *Recorded Sound*, i (1961–2), 242
E. Sackville-West: 'Schnabel', *Recorded Sound*, i (1961–2), 40
——: '[Moriz] Rosenthal', *Recorded Sound*, i (1961–2), 214
P. Saul: 'Busoniana', *Recorded Sound*, i (1961–2), 256
H. O. Spingel: 'Emil Gilels: Phänomen der Tasten', *Fono-Forum*, vi/1 (1961), 14
B. Gavoty: *Claudio Arrau* (Geneva, 1962)
M. Weaver: 'Interview with Claudio Arrau', *Piano Quarterly*, no.42 (1962–3), 18
W. Gieseking: *So wurde ich Pianist* (Wiesbaden, 1963) [with discography by I. Hajmassy]
H. C. Schonberg: *The Great Pianists* (New York, 1963)
C. Halski: *Ignacy Jan Paderewski* (London, 1964)
W. Heiles: *Rhythmic Nuance in Chopin's Performances Recorded by Moriz Rosenthal, Ignaz Friedman, and Ignaz Jan Paderewski* (diss., U. of Illinois, 1964)
H. P. Range: *Die Konzertpianisten der Gegenwart* (Lahr, 1964, rev. and enlarged 2/1966)
B. Siki: 'Dinu Lipatti', *Recorded Sound*, no.15 (July 1964), 232
L. Biancolli: 'Moriz Rosenthal, "Last of the Pianistic Titans"', *HiFi/Stereo Review*, xiv/2 (1965), 54
R. Fermoy: 'Cortot', *Recorded Sound*, no.16 (Oct 1965), 266
J. Kaiser: *Grosse Pianisten in unserer Zeit* (Munich, 1965, 5/1982; Eng. trans., 1971, with enlarged discography)
F. Merrick: 'Memories of Leschetizky', *Recorded Sound*, no.18 (April 1965), 335
M. Moiseiwitsch: *Moiseiwitsch: Biography of a Concert Pianist* (London, 1965)
G. Seldon-Goth: *Ferruccio Busoni: un profilo* (Florence, 1965), 25
'Horowitz, Vladimir', *CBY 1966*
'Rubinstein, Artur', *CBY 1966*
H. Ferguson: 'Myra Hess', *Recorded Sound*, no.24 (Oct 1966), 102
M. Barzetti: 'Alfredo Casella (1883–1947)', *Recorded Sound*, no.28 (Oct 1967), 235
J. Moore: 'George Copeland', *Recorded Sound*, no.25 (Jan 1967), 140
H. Stuckenschmidt: *Ferruccio Busoni: Zeittafel eines Europäers* (Zurich and Freiburg, 1967; Eng. trans., 1970)
N. Cardus: 'Pianist in Exile', *International Piano Library Bulletin*, ii/4 (1968), 6 [on Ignacy Friedman; with discography]

H. P. Range: *Ins Herz geschaut: Anekdoten um Pianisten unserer Zeit* (Schauenburg, 1969)
I. Schwerke: 'Francis Planté: Patriarch of the Piano', *Recorded Sound*, no.35 (July 1969), 474
J.-J. Eigeldinger: *Chopin vu par ses élèves* (Neuchâtel, 1970, rev. 2/1979; Eng. trans., 1987, as *Chopin: Pianist and Teacher*)
D. Matthews: 'Edwin Fischer', *Recorded Sound*, no.39 (July 1970), 649
J. Wills: 'Egon Petri', *Recorded Sound*, no.39 (July 1970), 639
G. Sherman: 'Josef Lhévinne', *Recorded Sound*, no.44 (Oct 1971), 784
E. Barnett: 'An Annotated Translation of Moriz Rosenthal's "Franz Liszt, Memories and Reflections"', *CMc*, no.13 (1972), 29
F. C. Campbell: 'Lhévinne, Josef', *DAB*
Artur Rubinstein: *My Young Years* (London, 1973)
R. Threlfall: *Sergei Rachmaninoff: his Life and Times* (London, 1973), 46
Piano Quarterly, no.84 (1973–4) [special issue on Artur Schnabel]
R. R. Gerig: *Famous Pianists and their Technique* (Washington, DC, 1974)
Recorded Sound, no.63–4 (July–Oct 1976) [special issue on Clara Haskil]
A. Brendel: *Musical Thoughts and Afterthoughts* (London, 1976, rev. 2/1982)
R. K. Wallace: *A Century of Music-making: the Lives of Josef and Rosina Lhévinne* (Bloomington, Ind., 1976)
B. Gavoty: *Alfred Cortot* (Paris, 1977)
F. Lamond: 'Memories of Anton Rubinstein', *Recorded Sound*, no.65 (Jan 1977), 635
A. Walker: 'Frederic Lamond: 1868–1948', *Recorded Sound*, no.65 (Jan 1977), 636
J. Horowitz: 'Shura Cherkassky – a Pianist who Follows his Intuition', *New York Times* (2 April 1978)
L. Krebs: 'Eduard Erdmann, Pianist and Composer (1896–1958)', *Recorded Sound*, no.69 (Jan 1978), 762
D. Mayher: 'Fanny Davies', *Recorded Sound*, no.70–71 (April–July 1978), 776
B. Ott: *Liszt et la pédagogie du piano* (Issy-les-Moulineaux, 1978)
G. Payzant: *Glenn Gould* (London and Toronto, 1978)
Great Pianists Speak with Adele Marcus (Neptune, NJ, 1979)
Piano-Jahrbuch, i– (1980–)
'Who's Who of Pianists: Murray Perahia talks to Carola Grindea', *Piano Journal*, i/2 (1980), 7
E. Lipmann: *Artur Rubinstein ou l'amour de Chopin* (Paris, 1980)
J. Meyer-Josten: *Musiker im Gespräch: Claudio Arrau* (Frankfurt am Main, New York and London, 1980)
Artur Rubinstein: *My Many Years* (New York, 1980)
P. Susskind Pettler: 'Performers & Instruments: Clara Schumann's Recitals, 1832–50', *19th Century Music*, iv/1 (1980–81), 70
'Who's Who of Pianists: Vladimir Ashkenazy talks to Carola Grindea', *Piano Journal*, ii/6 (1981), 5
D. Calapai: 'Schnabel and Schubert', *Piano Journal*, ii/5 (1981), 7
E. Mach: *Great Pianists Speak for Themselves* (London, 1981)

J. Methuen-Campbell: *Chopin-playing: from the Composer to the Present Day* (London, 1981)
J. Meyer-Josten: *Musiker im Gespräch: Svjatoslav Richter* (Frankfurt am Main, New York and London, 1981)
B. Ogden and M. Kerr: *Virtuoso: the Story of John Ogden* (London, 1981)
C. Pangels: *Eugen d'Albert* (Zurich, 1981)
F. Schwarz: 'Emil Gilels', *Recorded Sound*, no.80 (July 1981), 1–76 [with discography]
'Perahia, Murray', *CBY 1982*
G. Kehler: *The Piano in Concert* (Metuchen, NJ, 1982)
H. P. Range: *Pianisten im Wandel der Zeit* (Lahr, 1982)
H. Sachs: *Virtuoso* (London, 1982)
D. Wilde: 'Solomon: an Appreciation', *Recorded Sound*, no.82 (July 1982), 1
A. Zamoyski: *Paderewski* (London, 1982) [with extensive bibliography]
'Who's Who of Pianists: Jorge Bolet talks to Carola Grindea', *Piano Journal*, iv/10 (1983), 5
J. Chissell: *Clara Schumann: a Dedicated Spirit* (London, 1983)
I. Harden: *Claudio Arrau: ein Interpretenportrait* (Frankfurt am Main, 1983)
M. Harrison: 'Arrau at 80', *Music and Musicians* (Feb 1983), 8
J. Horowitz: *Conversations with Arrau* (London, 1983)
J. Methuen-Campbell: 'Early Soviet Pianists and their Recordings', *Recorded Sound*, no.83 (Jan 1983), 1
G. Plaskin: *Horowitz: a Biography* (London and New York, 1983)
P. Rattalino: *Da Clementi a Pollini*: duecento anni con i grandi pianisti (Florence, 1983)
A. Walker: *Franz Liszt: the Virtuoso Years 1811–1847* (London, 1983)
M. Zilberquit: *Russia's Great Modern Pianists* (Neptune, NJ, 1983)
'Who's Who of Pianists: Shura Cherkassky talks to Carola Grindea', *Piano Journal*, v/13 (1984), 5
V. Ashkenazy and J. Parrott: *Ashkenazy: Beyond Frontiers* (London, 1984)
J. Cott: *Conversations with Glenn Gould* (Boston, 1984)
J. Hoskins: *Ignacy Jan Paderewski, 1860–1941: a Biographical Sketch and a Selective List of Reading Materials* (Washington, DC, 1984)
D. Pistone, ed.: *Sur les traces de Frédéric Chopin* (Paris, 1984)
D. Dubal: *The World of the Concert Pianist* (London, 1985)
W. Lyle: *A Dictionary of Pianists* (London, 1985)
H. Neuhaus: 'Sviatoslav Richter: a Profile', *Piano Journal*, vi/18 (1985), 15
N. Reich: *Clara Schumann: the Artist and the Woman* (London, 1985)
'Jorge Bolet Speaks Out', *Piano Quarterly*, xxxiv (1986), spr., 15
D. Elder: *Pianists at Play* (London, 1986)

BIBLIOGRAPHY

REPERTORY

General reference

C. Weitzmann: *Geschichte des Clavierspiels und der Clavierliteratur* (Stuttgart, 1823, rev. and enlarged, 2/1879)

W. Niemann: *Das Klavierbuch: Geschichte der Klaviermusik und ihrer Meister* (Leipzig, 1922)

H. Westerby: *The History of Pianoforte Music* (New York, 1924)

E. Blom: *The Romance of the Piano* (London, 1928)

D. F. Tovey: *Essays in Musical Analysis* (London, 1935–9)

A. Lockwood: *Notes on the Literature of the Piano* (Ann Arbor and London, 1940)

A. Loesser: *Men, Women and Pianos* (London, 1940)

W. Georgii: *Klaviermusik* (Zurich and Freiburg, 1941, 4/1965)

D. Brook: *Masters of the Keyboard* (London, 1946)

E. Hutcheson: *The Literature of the Piano* (New York, 1948, rev. 3/1964/*R*1974)

W. Georgii: *Klaviermusik* (Zurich, 1950)

J. Friskin and I. Freundlich: *Music for the Piano ... from 1580 to 1952* (New York, 1954/*R*1973)

J. Gillespie: *Five Centuries of Keyboard Music* (Belmont, Calif., 1965/*R*1972)

F. E. Kirby: *A Short History of Keyboard Music* (New York, 1966)

K. Wolters: *Handbuch der Klavierliteratur*, i (Zurich and Freiburg, 1967)

M. Hinson: *Keyboard Bibliography* (Cincinnati, 1968)

D. Matthews, ed.: *Keyboard Music* (London, 1972)

M. Hinson: *Guide to the Pianist's Repertoire*, ed. I. Freundlich (Bloomington, Ind., 1973; suppl., 1980) [comprehensive bibliography]

D. Gill, ed.: *The Book of the Piano* (Oxford, 1981)

M. Hinson: *Music for Piano and Orchestra: an Annotated Guide* (Bloomington, Ind., 1981)

J. Frances: *A Compendium of Piano Concertos* (Hawthorne, NJ,1981)

C. McGraw: *Piano Duet Repertoire* (Bloomington, Ind., 1981)

R. Fuszek: *Piano Music in Collections: an Index* (Detroit, 1982)

P. Griffiths: *British Music Catalogue 1945–81*, i: *Works for Piano* (London, 1983)

W. Phemister: *American Piano Concertos: a Bibliography* (Detroit, 1985)

Specific studies

E. J. Dent: 'The Pianoforte and its Influence on Modern Music', *MQ*, ii (1916), 271

A. Cortot: *La musique française de piano* (Paris, 1930–48; i–ii, Eng. trans., 1932)

C. Parrish: *The Early Piano and its Influence on Keyboard Technique and Composition in the Eighteenth Century* (diss., Harvard U., 1939)

J. F. Russell: 'Mozart and the Pianoforte', *MR*, i (1940), 226

N. Broder: 'Mozart and the Clavier', *MQ*, xxvii (1941), 422

E. Reeser: *De zonen van Bach* (Amsterdam, 1941; Eng. trans., 1946)

W. Apel: *Masters of the Keyboard* (Cambridge, Mass., 1947)

C. Parrish: 'Haydn and the Piano', *JAMS*, i/1 (1948), 27

J. Kirkpatrick: 'American Piano Music: 1900–1950', *MTNA Proceedings*, xliv (1950), 35
F. H. Garvin: *The Beginning of the Romantic Piano Concerto* (New York, 1952)
D. Stone: *The Italian Sonata for Harpsichord and Pianoforte in the Eighteenth Century (1730–90)* (diss., Harvard U., 1952)
A. G. Hess: 'The Transition from Harpsichord to Piano', *GSJ*, vi (1953), 75
K. Dale: *Nineteenth Century Piano Music* (London, 1954) [foreword by Myra Hess]
H. F. Wolf: *The 20th Century Piano Sonata* (diss., Boston U., 1957)
E. Blom: 'The Prophecies of Dussek', *Classics Major and Minor* (London, 1958), 88
N. Demuth: *French Piano Music* (London, 1958)
T. L. Fritz: *The Development of Russian Piano Music as Seen in the Literature of Mussorgsky, Rachmaninoff, Scriabin, and Prokofiev* (diss., U. of Southern California, 1959)
P. F. Ganz: *The Development of the Etude for Pianoforte* (diss., Northwestern U., 1960)
J. Lade: 'Modern Composers and the Harpsichord', *The Consort*, xix (1962), 128
W. S. Newman: *The Sonata in the Classic Era* (Chapel Hill, 1963, rev. 3/1983)
E. Badura-Skoda: 'Textural Problems in Masterpieces of the Eighteenth and Nineteenth Centuries', *MQ*, li (1965), 301
T. A. Brown: *The Aesthetics of Robert Schumann in Relation to his Piano Music 1830–1840* (diss., U. of Wisconsin, 1965)
L. D. Stein: *The Performance of Twelve-tone and Serial Music for the Piano* (diss., U. of Southern California, 1965)
M. J. E. Brown: 'Towards an Edition of the Pianoforte Sonatas', *Essays on Schubert* (New York, 1966), 197
K. Heuschneider: *The Piano Sonata in the 18th Century in Italy* (Cape Town, 1966)
D. L. Arlton: *American Piano Sonatas of the Twentieth Century: Selective Analysis and Annotated Index* (diss., Columbia U., 1968)
M. K. Ellis: *The French Piano Character Piece of the Nineteenth and Early Twentieth Centuries* (diss., Indiana U., 1969)
E. Glusman: *The Early Nineteenth-century Lyric Piano Piece* (diss., Columbia U., 1969)
W. S. Newman: *The Sonata since Beethoven* (Chapel Hill, 1969, rev. 3/1983)
K. Michałowski: *Bibliografia chopinowska 1849–1969* (Kraków, 1970)
W. S. Newman: 'Beethoven's Pianos versus his Piano Ideals', *JAMS*, xxiii (1970), 484
H. Truscott: 'The Piano Music – I', *The Beethoven Companion*, ed. D. Arnold and N. Fortune (London, 1971)
K. Michałowski: 'Bibliografia chopinowska 1970–1973', *Rocznik chopinowski*, ix (1975), 121–75
E. Badura-Skoda: 'Prolegomena to a History of the Viennese Fortepiano', *Israel Studies in Musicology*, ii (1980), 77

BIBLIOGRAPHY

L. Powell: *A History of Spanish Piano Music* (Bloomington, Ind., 1980)
C. Maxwell, ed.: *Haydn: Solo Piano Literature* (Boulder, Colorado, 1983)
K. Wolff: *Masters of the Keyboard: Individual Style Elements in the Piano Music of Bach, Haydn, Mozart, Beethoven, and Schubert* (Bloomington, Ind., 1983)
P. Fallows-Hammond: *Three Hundred Years at the Keyboard: a Piano Sourcebook from Bach to the Moderns* (Berkeley, Calif, 1984)
J. and A. Gillespie: *A Bibliography of Nineteenth-century American Piano Music* (Westport, Conn., 1984)
M. Jost: *Practice, Interpretation, Performance: Aspects of the Pianoforte Repertoire* (Melbourne, 1984)
C. Maxwell, ed.: *Schumann: Solo Piano Literature* (Boulder, Colorado, 1984)

Index